SpringerBriefs in Electrical and Computer Engineering

T0210544

More information about this series at http://www.springer.com/series/10059

Jianhua Lu • Xiaoming Tao • Ning Ge

Structural Processing for Wireless Communications

Foreword by Khaled B. Letaief, FIEEE, FHKIE

 Springer

Jianhua Lu
Department of Electronic Engineering
Tsinghua University
Beijing, China

Xiaoming Tao
Department of Electronic Engineering
Tsinghua University
Beijing, China

Ning Ge
Department of Electronic Engineering
Tsinghua University
Beijing, China

ISSN 2191-8112   ISSN 2191-8120   (electronic)
SpringerBriefs in Electrical and Computer Engineering
ISBN 978-3-319-15710-8   ISBN 978-3-319-15711-5   (eBook)
DOI 10.1007/978-3-319-15711-5

Library of Congress Control Number: 2015932994

Springer Cham Heidelberg New York Dordrecht London

Printed on acid-free paper

Springer International Publishing AG Switzerland is part of Springer Science+Business Media (www.
springer.com)

Recommended by Sherman Shen

Foreword

As wireless communication systems continue to grow, their designs are becoming increasingly complex, whereas the problem of identifying better system designs is posing severe challenges not only to the academia but also to the industry. Specifically, as the number of mobile subscribers has been increasing at an incredible speed in recent years, services have rapidly diversified to meet widely divergent user needs and a variety of stringent requirements. This exacerbates the difficulty of the problem, since analyzing the complexity of systems where various services are involved is very difficult, as such systems cannot be decomposed into sub-systems that exhibit linear properties. Likewise, interferences caused by the coexistence of different elements in the same physical channel makes the wireless environment harsher than ever, highlighting the uncertainty issues in wireless transmission. As a result, the classic communication theory has reached its limits.

This brief presents an alternative viewpoint on processing technologies for wireless communications based on recent findings. The structure perspective presented acts as a lever in enabling emerging processing technologies and helps to cope with the aforementioned challenges. Unlike classic processing methods that are mainly element based (they operate on elements such as bits or samples), in this brief, wireless multimedia communication, channel coding and pre-coding methods are designed with different structures as intermediate processing units. Based upon mathematical analysis and simulation results, the proposed technologies are convincingly shown with promising performance in broadband wireless multimedia systems. I personally believe that the idea of structural processing is unique and innovative and hope this brief may boost the development of new theories and technologies with this perspective, which would likely open up a new trend in wireless communications research.

With a great pleasure, I highly recommend this brief to everyone in the area of information and communication technology.

Hong Kong, China   Khaled B. Letaief

Preface

In the early twentieth century, the revolution in wireless communication began with the pioneering attempt of long-distance radio communications by Guglielmo Marconi, who performed the famous *transatlantic* transmissions in 1901 from Newfoundland, Canada, to Cornwall, Britain. Other great pioneers include Samuel Morse (telegraph), Alexander G. Bell (telephone), as well as Edwin Armstrong (radio).

The greatest breakthrough in theory emerged with the publication of Claude Shannon's theorems in 1948. These theorems underlay the performance limits in encoded transmission, which catalyzed the development of a new area in science currently known as *information theory*. Under the effective guidance from Shannon's theorems, processing technologies have been designed for compressing data, increasing link speed, reducing transmission errors and so on. There have been much efforts to push the performance closer to the Shannon limit, but challenges continue to exist when facing the ever-growing complexity of systems. Indeed, one necessary prerequisite of applying Shannon's theorems is that the source and the channel models should both possess independent identical distributions (*i.i.d.*). When the complexity and uncertainty issues in wireless systems get so severe that this assumption deviates too greatly from the truth, the classic processing technologies will meet difficulties in achieving high transmission performance.

To address the problems caused by complexity and uncertainty, we hereby attempt to develop an alternative processing method called *structural processing*. Instead of operating on bits or samples based on the *i.i.d.* model, structural processing tries to handle grouped bits or samples, namely, structures. The study of structural processing should be traced back to 18 years ago when I started my research as a PhD student, under the supervision of Prof. Khaled Ben Letaief and Prof. Ming L. Liou at The Hong Kong University of Science and Technology in 1996. At the very beginning, we found that the impact on the perceptive image quality caused by bit errors with a fixed error rate in the transmission of JPEG files varied greatly across different trial runs. The uncertainty of the effect of the errors was mainly due to the complexity of the underlying data structure being corrupted by the errors. By learning the data structure of the JPEG format, the concept called *minimum data unit (MDU)* was proposed, which grouped the related encoded bits representing an image block. Once MDU is transmitted as an entity, its error rate, i.e., $E_r(MDU)$, may act as a bridge between received image quality and bit error rate, or, BER, performance. This is in fact an initiation to have the idea of structural processing. In the meantime, a block shuffling scheme was proposed, together by Mr. I. P. Chan, Prof. J. C.-I. Chuang and myself, as a structural processing method in order to enhance the error resilience of image transmission. Part of this work is introduced in Chap. 3.

Later on, Prof. Ning Ge, Dr. Liuguo Yin, Dr. Yukui Pei and Dr. Xiaoming Tao joined our group in Tsinghua University, continuing the research on structural processing. It was noted that IP over SDH and IP over ATM faced a similar issue regarding the high variability of the effect of errors as that which MDU was proposed to solve in data communications. We studied the framing procedure on the data link layer which prevented IP structure corruption from transmission errors to protect the integrity of variable-length IP packets. This further generalized our research on wireless transmission based on structures. Some important ideas are included in Chap. 2.

Channel coding is an important part of a communication system to combat errors in transmission. It is well known that conventional LDPC code design is usually a bit-based random search for a coding matrix, where the number of bit-element combinations is usually huge. To address this complicated matrix optimization problem, we proposed a structured method based on Galois field sub-matrices, which decomposed the huge matrix design problem into a succession of sub-matrix design problems. This structured coding design is another demonstration of structural processing, which is introduced in Chap. 4.

In 2007, Mr. Weiliang Zeng started his study in our group. We carried on a collaborative research on pre-coding design together with Prof. Chengshan Xiao from Missouri University. Although classic pre-coding design may achieve optimal system capacity, yet it is nearly infeasible for practical implementations, for its derivation is based on the assumption of Gaussian signals, which are both unbounded and undetectable. Considering a finite-alphabet constellation structure of the modulation signal, we proposed a structured pre-coding design with a two-step iterative optimization algorithm. Benefiting from the merit of structural processing, the developed algorithm may provide a performance gain that brings the system capacity close to the optimum with acceptable computing complexity. Part of this work is presented in Chap. 5.

Since 2010, Mr. Yang Li, Mr. Yipeng Sun and Mr. Shaoyang Li have joined us to conduct research of structural processing towards their respective PhDs. Instead of finding and utilizing structures constructed by correlated bits, we have then focused on a structural perspective inspired by the cognition of human brains. Specifically, images and videos are modeled based on high-level perceptions and represented with structural decomposition. Consequently, dictionary learning methods for image representation and model-based approaches for face video communications are studied with improved transmission efficiency and perceptive quality. As a matter of fact, these technologies deal with the non-i.i.d. sources and channels without simple transformation or approximation to i.i.d. ones. Some interesting results are briefly presented in Chap. 6.

In this brief, we have therefore assembled exemplar works from our 18 years of team research, where methods of source coding, channel coding and pre-coding are presented with structure perspectives. Specifically, the concept of MDU is introduced to help redesign the compressed wireless multimedia data, so as to significantly curb the impact of both random and burst errors. Likewise, a gradual construction scheme of structured LDPC codes is found, having near-optimal error-correction performance but much lower complexity than the conventional LDPC coding. Moreover, pre-coding for high-dimension constellation consisting of the basic QPSK structure is designed, achieving near-optimal capacity in harsh wireless environments. Further studies on information representation, such as dictionary learning and model-based video coding, are introduced as potential interesting areas to the development of advanced structural processing technologies for future wireless communications.

This brief only showcases some recent processing technologies with structural perspective. Hence, more research work is encouraged to follow up. Although the introduced technologies belong to different parts of the wireless system, the theme is unified: *structure*. It is the structure that enables the processing complexity of non-i.i.d. systems may be decomposed, facilitating further studies on more general methods to tackle the complexity and uncertainty issues in modern wireless communication systems. We hope that this brief may provide alternative ideas to the researchers and also be used as a reference for both post and undergraduate students who major in wireless communication, information theory or related areas. Our research thus far constitutes preliminary explorations into their respective topics, with inevitable flaws and limitations. Much work remains to be done toward the eventual formulation of a comprehensive theoretical framework.

In the meantime of writing this brief, some members in our group in Tsinghua University participated in preparing the materials and provided valuable assistance. Specifically, Dr. Linhao Dong assisted in editing Chap. 1; Dr. Yukui Pei provided simulation results and helped edit Chap. 4; Mr. Shaoyang Li prepared part of Chap. 2 and helped edit Chap. 6; Dr. Rui Shi helped edit Chap. 3 and part of Chap. 4; Mr. Hongliang Mao helped complete Chap. 5; and Mr. Yang Li helped improve the quality of presentation of the brief. We highly appreciate their excellent work.

Beijing, China   Jianhua Lu
January 2015

Acknowledgements

This work was supported in part by the National Basic Research Projects of China (973), entitled "Fundamental Research on Theory of Smart Collaborative Broadband Wireless Networks" (Grant No. 2013CB329000), "Fundamental Research on Multi-Domain Collaboration for Broadband Wireless Communications" (Grant No. 2007CB310600) and "Research on High Performance Algorithms for Multimedia Communications and System Integration Technology" (Grant No. 1998030406), together with the Key Projects of National Natural Science Foundation of China, entitled "Research on Key Technologies in Deep Space Communications" (Grant No. 60532070), "Next Generation Wireless Multimedia Network System, Theory and Applications" (Grant No. 60132010) and "Theory and Method of Coding and Modulation for High Speed Data Transmission in Aerospace Networks" (Grant No. 61132002), as well as the Projects for Innovative Research Group of National Science Foundation of China under grants No. 61321061 and 61021001.

Also sincere thanks to Tsinghua National Laboratory for Information Science and Technology, Department of Electronic Engineering and School of Aerospace Engineering of Tsinghua University for their grateful support. The authors highly appreciate contributions from all team members of Wireless Multimedia Communications Lab, Tsinghua University.

Contents

Acronyms

AWGN	additive white Gaussian noise
BER	bit error rate
bpp	bit per pixel
BPSK	binary phase-shift keying
CSCC	combined source-channel coding
CSI	channel state information
DCT	discrete cosine transform
DPC	dirty paper coding
DPCM	differential pulse-code modulation
EM	electromagnetic (wave/field)
FEC	forward error correction
GF	Galois field
HEVC	high efficiency video coding
i.i.d.	independent identical distribution
IP	Internet protocol
JPEG	joint picture expert group
LDPC	low-density parity-check (code)
MBVC	model-based video coding
MDU	minimum data unit
MIMO	multiple-input and multiple-output
MMSE	minimum mean square error
MSR	multi-sample sparse representation
OFDM	orthogonal frequency-division multiplexing
PB	picture block
PCA	principal component analysis
PSNR	peak signal-to-noise ratio
QAM	quadrature amplitude modulation
QPSK	quadrature phase shift keying
R-D	rate-distortion
ROI	region-of-interest
RTP	real time transport protocol

SER	symbol error rate
SNR	signal-to-noise ratio
SSIM	structural similarity index measure
SVD	singular value decomposition
VLC	variable length coding
WER	word error rate

Chapter 1
Revisiting Wireless Communications

1.1 Overview

There is no doubt that wireless communication technology has vastly improved the quality of our daily life. For short-range communications, Wi-Fi (IEEE 802.11a/b/g/n) has been widely incorporated in many gadgets such as laptops, tablet computers, and digital cameras for efficient Internet access required by many interesting applications. For telecommunication services, smart phones have dominated the mobile market recently with the full deployment of the 3G/4G mobile networks. Integrating both short-range and cellular wireless transceivers, smart phones provide a dynamic platform for various services such as instant messaging, Internet access, business applications, on-line gaming, multimedia services, video-conferencing, and of course, voice telephony. Consequently, more system capacity in wireless networks is demanded. Since its launch in 2010 [1], 4 G network has become the technology where most place their hope for the continued enlargement of current network capacities in the coming years, offering a target data rate up to 1 Gbps [2].

Apart from telecommunication services, wireless communication is also utilized in other fields. In astronomical research, wireless communication plays a unique part for building links between ground stations and thousands of artificial satellites, space probes, and spacecrafts in outer space. Usually, the ground station sends control signals to the unmanned spacecraft, and the spacecraft returns data in various forms such as images and waveforms to the ground station. In military and modern warfare, the role of wireless communications is irreplaceable. For example, precision-guided munitions, such as cruise missiles, are largely equipped in many troops to maximize the operational efficiency while minimizing civilian casualties. Airborne Warning and Control System (AWACS) and satellite communication systems are also introduced for command and control.

© The Author(s) 2015
J. Lu et al., *Structural Processing for Wireless Communications*, SpringerBriefs in Electrical and Computer Engineering, DOI 10.1007/978-3-319-15711-5_1

In a word, wireless communication has long become an essential part in every facet of our lives, and will continue to be one of dominant technologies in the following decades.

While an astonishing number of different systems have been invented and deployed, the basic principle is mostly same. Namely, original information passes through a series of processing units, followed by transformation to electromagnetic (EM) waves for delivery. In the early years of communications, scientists and engineers started to realize the necessity of measuring the quantity of information in transmissions. In 1948 [3], Claude Shannon published his theorems for the communications of discrete messages, based on previous works by Harry Nyquist and Ralph Hartley. Now these theorems are commonly known as *source coding theorem*, *rate-distortion theorem* and *noisy-channel coding theorem*, which are briefly described as follows:

- **Source coding theorem [4]:** if N random variables follow independent identical distribution (i.i.d.) $p(x)$ with entropy $H(X)$, the minimum expected length L in bits to represent each variable satisfies $L \to H(X)$ as $N \to \infty$. On the contrary, if they are compressed into bits that are fewer than $H(X)$ per variable on average, it is virtually certain that information will be lost.
- **Rate-distortion theorem:** for an i.i.d. source X with distribution $p(x)$ and bounded distortion function $d(x, \hat{x})$, the minimal number of bits per data sample, measured as rate R, can be determined, where the input signal can be approximately reconstructed at the receiver without exceeding a given distortion D. R is equal to the associated information rate distortion function, which is

$$R(D) = R^{(I)}(D) = \min I(X; \hat{X}), \tag{1.1}$$

where $I(X; \hat{X})$ is the mutual information between X and $\hat{X}$, and the minimization is over all conditional distribution $p(\hat{x}|x)$ satisfying the expected distortion constraint $d(x, \hat{x}) \leq D$.
- **Noisy-channel coding theorem [5]:** for a discrete memoryless channel, it is possible to communicate digital information nearly error-free at any rate blow the channel capacity.

These theorems draw quantitative boundaries on data compression and reliable transmission. Thanks to them, a 60-year prosperity of wireless communications has been witnessed.

From a system point of view, Fig. 1.1 shows the block diagram of a general wireless communication system in the unidirectional transmission. If the input is analog, it is usually sampled and quantized. This procedure, called sampling, serves as a very important step to convert the analog signals to digitalized source information. If the input is digital, it may be converted to a certain binary sequence directly. However, the raw binary data is typically too massive to be stored or transmitted; therefore, lossless or lossy compression is usually applied to reduce the size of the data meanwhile maintaining an acceptable level of fidelity.

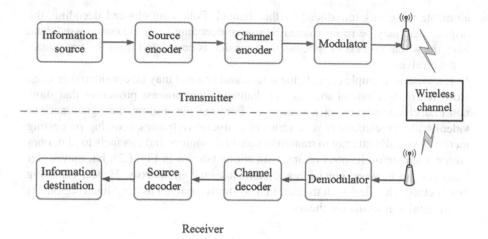

Fig. 1.1 Diagram of a general wireless communication system, where the transmission and reception can be reciprocal

At this stage, the compressed output is stored as bit sequences in a storage medium. These bit sequences contain the principal information of the original data. To combat the errors from transmission over noisy channels, channel coding or forward error correction (FEC) is applied, where a code-word is generated in a redundant way by adding checking bits. If enough checking bits are used, it is possible to decode the original sequence without error. This is de facto the main point of noisy-channel coding theorem. Usually, the design of the code determines the performance of error correction. General channel coding methods may be categorized into block codes and convolutional codes, where block codes are memoryless and convolutional codes are typically designed with finite-state machines. Amongst these, low density parity check (LDPC) codes may exhibit performance approaching the Shannon limit [6] in additive white Gaussian noise (AWGN) channel transmission. However, the large scale of its random-like generator matrix significantly increases the complexity of design. Therefore, the problem of finding a code with good performance as well as low complexity always attracts researchers' attention. After channel coding, the bit sequences are mapped to symbols, and these symbols are converted to signals with digital modulation. After mixing and amplification, a continuous waveform is generated and then emitted from an antenna.

At the receiver, the waveform is firstly picked up by an antenna before passing through the low-noise amplifier and the RF downconvertor, and then passes blocks such as matched filter, sampler, and decision devices inside a demodulator. After demodulation, the waveform is converted back to bit sequence, which may contain errors caused by transmission impairments such as multi-path fading, interferences, carrier frequency offset, noise, and so on, in propagation. To recover the transmitted bit sequences, the decoder utilizes either hard or soft decision algorithms to eliminate the errors introduced by the channel. Following channel decoding, the original data may be reconstructed by source decoding from the compressed data. Now the whole process of wireless transmission is completed from the transmitter to the receiver.

If a system is simple enough, the source and channel may be considered as i.i.d.. However, many kinds of sources and channels often possess properties that show non-i.i.d. characteristics, such as memory. For example, multimedia signals such as videos have correlations cross a series of consecutive frames. Existing processing methods typically attempt to transform non-i.i.d. sources and channels to i.i.d. ones with certain simplifications or approximations (shown in Fig. 1.2), but may suffer from conflict between precision and computational complexity. Hence, exploring new methods that deal with the non-i.i.d. systems, denoted as gray-lined regions, is meaningful in information theory.

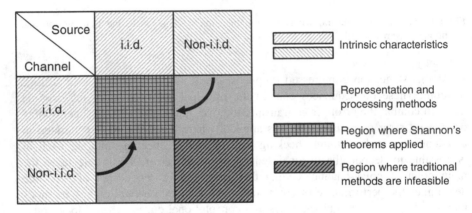

Fig. 1.2 An illustration of the relationship between processing methods and the properties of source and channel, where the *white-lined* and *gray* regions represent the intrinsic characteristics of source and channel and corresponding processing methods, respectively. Conventional methods likely transform non-i.i.d. source or channel to i.i.d. ones, where Shannon's theorems can be readily applied

In the following sections, we will discuss the challenges in wireless communications through an analysis of the issues arising from complexity and uncertainty.

1.2 Challenges in Emerging Wireless Systems

1.2.1 Difficulties in Wireless Multimedia Communications

The early 2000s have witnessed the breakthrough of the development in mobile communications. In particular, two 3G standards, universal mobile telecommunications system (UMTS) [7] and code division multiple access 2000 (CDMA2000) [8]

were deployed. One of the outstanding features of the 3G standards is multimedia service support. The latest addition to the 3G standards, evolved high-speed packet access (HSPA+), may provide an average data rate of approximate 85 Mbps in downlink (Release 9) [9]. Meanwhile, the rising popularity of smart phones supporting high-speed mobile network standards fuels customers' increasing needs for wireless multimedia services.

It is noted that, early research on wireless communications did not distinguish multimedia communication from conventional data communication; they are simply considered to be a simple combination of two individual parts: multimedia and communication. The differences and correlations among bits are usually neglected in the design of multimedia transmission systems.

Under the assumption of ideal error-free transmission, Shannon's rate-distortion theorem is an efficient guide to the design of source coders for multimedia signals. With such design, an optimal trade-off between reconstruction quality and compression efficiency may be achieved [10]. However, in practical wireless transmission, the compressed data are rather sensitive to transmission errors due to the inter-correlation property. When transmission errors exceed a certain threshold and cause significant data to become corrupted, the quality of multimedia reconstruction may be drastically decreased, leading to a *cliff effect* [11], which does not follow the rate-distortion theorem. The method of dealing with this cliff effect in multimedia communication is one of the great challenges in contemporary research [11].

1.2.2 Complex Interferences in Harsh Radio Environments

Differing from wireline communications, interferences are present whenever frequency resources are shared by multiple users. The current trend of development in wireless sees a shrinkage of individual cell coverage, and a growth of the number of antennas. For instance, Long-Term Evolution (LTE) system applies techniques such as multiple-input and multiple-output (MIMO) and orthogonal frequency-division multiplexing (OFDM) to meet the growing needs in data services from new consumers with increased system capacity [12]. However, the interferences from multiple antennas and users are often difficult to be removed completely, preventing improvement in system capacity from reaching theoretically expected effectiveness. Furthermore, as the numbers of nodes and antennas increase, more spectrum overhead is required for channel estimation and feedback links carrying channel state information (CSI).

Likewise, modern military communications face harsh EM environments as well. Tactical schemes such as frequency hopping, radar detection, and EM attacking introduce dense and dynamic overlaps at spatial, time, and frequency domains. As a result, problems including inconsistent transmission, high bit-error rate (BER), and low spectral efficiency are often encountered.

1.2.3   The Challenge of Distance in Deep Space
Communications

Since the launch of the first artificial satellite Sputnik 1 by the Former Soviet
Union in 1957, thousands of satellites have been launched to provide services in
communication, navigation, remote sensing, weather forecast, etc. Modern wireless
technologies are able to cope with communications over long distances (e.g., low
Earth orbit (LEO) satellites at about 1,000 km and geosynchronous orbit (GEO)
satellites at about 36,000 km). Certain man-made objects such as space probes
and spacecrafts launched for astronomical research travel an extraordinarily long
distances from the earth into deep space far beyond this range. Since path loss (PL)
of wireless signals in free space is in proportion to the square of the transmission
distance [13], the received signal-to-noise ratio (SNR) of the transmitted signals
in deep space communication may be very low. For instance, lunar probes need to
combat 21 dB more attenuation in transmission than the GEO satellites [14]. In such
cases, noise may be dominant in the received signal, resulting in a severe issue of
uncertainty to robust reception.

To enhance the capability of deep space communications, various approaches
have been studied and successfully applied, including enlarging the aperture of
antennas, increasing transmission power, reducing thermal noise at the receiv-
ing end, increasing the frequencies of carriers, and designing efficient channel
codes, etc.. Nevertheless, as the transmission distance becomes longer and longer,
the effects of communication uncertainty will be more and more severe, which will
continue to reduce the maximum achievable transmission rate. For example, by June
19, 2014, Voyager 1 has traveled around 127.34 AU (1.9×10^{10} km) from the earth,
at the boundary between heliopause and bow shock [15].

Also, deep space communications may encounter complex EM environments due
to burst interferences from solar wind, plasma, geomagnetic storm and other EM
waves, which likely cause unpredictable burst errors during data transmission over
space. In 1997, the Jet Propulsion Laboratory (JPL)-launched robotic space probe
NEAR-Shoemaker suffered from fierce solar scintillation at a place 3.17 AU away
from earth, resulting in a frame-correction rate down to 3 % [16].

Although hundreds of pictures in detail of other planets of the solar system had
been transmitted back to Earth during the past 30 years, the data rate is currently
limited to the level of Kbps, and will be further shrunk with the fast increase of
spacecraft range in the future.

These challenges in the aforementioned emerging systems have obstructed the
application progress of wireless communications.

1.3 The Issues of Complexity and Uncertainty

1.3.1 The Ever-Increasing Complexity in Wireless Systems

Generally, the analysis of the complexity in wireless communications depends on two crucial system properties: the decomposability of systems and the stationary ergodicity of processes throughout systems.

Decomposability here means that a system can be decomposed into several simple and mutually independent sub-systems for individual study and analysis, by which the overall system performance may be obtained accordingly [17]. One classic example is a linear time-invariant (LTI) system, whose output may be described as the weighted sum of a series of impulse responses, thus a firm theoretical foundation may be established for its system design and optimization. Similarly, classic wireless communication theory thoroughly adopts this idea. For instance, the separate source-channel coding theorem by Shannon indicates that, if the source follows a certain i.i.d., and the channel is stationary, designing source and channel coding schemes separately can achieve the equivalent optimality as joint source-channel coding [5]. Base on this idea, the complexity in design and application of communication processing algorithm may be significantly reduced. However, separate coding theorem is not always applicable. The multiple access channel is a typical exception [5].

Assuming that two sources U and V send binary sequences to one destination simultaneously via an abstract multiple access channel. The joint distribution $p(u, v)$ is $\{1/3, 1/3, 0, 1/3\}$ for $\{(0, 0), (0, 1), (1, 0), (1, 1)\}$ respectively. The multiple access channel is discrete additive noiseless, and the output of the destination is the sum of the sources. In this case, the capacity region of the multiple access channel does not intersect with the Slepian-Wolf rate region of the distributed source encoding [5]. The source with Slepian-Wolf encoding to achieve the best source encoding performance will not use the multiple access channel efficiently, because the capacity of the multiple access channel increases with the correlation between the inputs. But if the source and the channel are jointly considered, a simple scheme that directly sends the two sources' data to the channel will reach the destination without error.

In practical wireless networks, we have to face the difficulties arisen from indecomposable complexity when calculating the channel capacity. For a long time, researchers had attempted to analyze the capacity in wireless networks. However, closed-form expressions or upper/lower capacity bounds have only been found for a few particular network models. In 1971, R. Ahlswede derived the capacity region of multiple access channels (many-to-one) [18]. Regarded as the reciprocal case of the multiple access channel, the capacity region of the broadcast channel (one-to-many) was obtained afterwards [19]. E. C. van der Meulen and T. Cover further derived the capacity bounds for three-terminal channels [20, 21], which are considered as one of the most influential outcomes in channel capacity.

Results for more general models still remain unknown. For example, although Shannon defined the duplex channel, we still have not yet found the explicit expression of capacity for this channel. Due to the fact that one wireless network cannot be decomposed into several individual point-to-point links under the existence of coupling, even the capacity of the two-user Gaussian interference channel remains open. The research in channel capacity is still in progress.

The stationary ergodicity of the signal processes in wireless communications is widely utilized to simplify optimization. In other words, assumption of stationary ergodicity ensures optimality in conventional communication systems. Taking the resource allocation problem as an example, classic methods allocate available channel or network bandwidth resources (spectrum, time, spatial resources, etc.) based on the need of data services [5], and often simplify complex models of services, channels and networks with several parameters such as bandwidth and delay. In fact, the common foundation of these methods is the stationary ergodicity of the processes of services and channels, such that the characteristic parameters may be collected from the statistical averages of a few samples during a unit time. As a result, the models built accordingly are only suitable for a single pair of stationary link and service, since their optimization only focus on individual samples.

However, it must be noted that the requirements on bandwidth, delay, and error rate control of mobile multimedia services are very different from those of conventional data communication services. For example, the rule in time-space variation of end-to-end quality of service (QoS), and the characteristics of group distribution in different time and space scales make multi-service oriented resource allocation as a difficult multi-parameter dynamic optimization problem. Very often, such problem is NP-hard in terms of computational complexity. Moreover, it is very unlikely for these kinds of services to possess stationary ergodicity so that they cannot be optimized simply with statistical averages. Therefore, instantaneous optimal results should be found in accordance with time-varying services and resources, making the optimization problem very hard to solve.

1.3.2   Uncertainty in Wireless Communications

The core of communication tasks is to tackle the uncertainty and eliminate its effects on information transmission. Shannon proposed entropy as a measurement of uncertainty. According to Shannon's definition, the entropy H of a random variable X can be written as:

$$H(X) = -\sum_{x \in \mathfrak{X}} p(x) \log p(x), \tag{1.2}$$

where $\mathfrak{X}$ is the value space of x, with $p(x)$ as its probability density function (PDF), i.e., $p(x) = \Pr\{X = x\}, x \in \mathfrak{X}$. Uncertainty in wireless communications include both source and channel uncertainty. Source uncertainty is expressed by the source entropy, which draws the lower bound on compression coding rate.

Shannon's source coding theorem [5] suggests that, when the source $X^n = \{X_1, \ldots, X_n\}$ follows i.i.d. (assuming each element of X^n has the same distribution with random variable Q), and as long as the compression coding rate R_S is under $R_S > H(Q)$, where $H(\cdot)$ denotes the entropy, it is possible to recover X^n with an arbitrary small error rate. The prerequisite of i.i.d. ensures that $2^{nH(Q)}$ sequences can represent all possible source sequences with a large probability. In other words, Shannon's coding theorem is applicable to stationary ergodic sources, but not to most of the multimedia sources that are neither stationary nor ergodic.

Uncertainty in channels usually comes from some factors such as noise, interference, channel fading, etc., and is measured by the channel capacity. For instance, the normalized capacity of AWGN channel can be written as:

$$C = \max_{p(x)} I(X;Y) = \frac{1}{2} \log\left(1 + \frac{P}{\sigma_0^2}\right), \tag{1.3}$$

where $p(x)$ is the probability density of input sources, and $I(X;Y)$ the average mutual information between the input X and output Y. P is defined as the average power of received symbols, and σ_0^2 the average noise power. Uncertainty introduced from the channel may be, in certain situations, removed by the use of error correction codes. The noisy-channel coding theorem points out that, when the transmit rate is less than the channel capacity, there must exist a certain error correction method to make the decoding error converge to zero [5]. Note that the channel here is assumed to be a stationary random process with ergodicity, and the channel code must be long enough. However, for wireless communications with mobility, time-varying fading is unlikely to exhibit stationary ergodicity.

1.3.3   The Issues of Complex Uncertainty

As complexity and uncertainty coexist in a system, the issue of complex uncertainty arises.

Figure 1.3 depicts a general analytical model of wireless communication systems. As shown, a non-stationary random sequence represents the information source, such as multimedia data after compression. Likewise, a random process of time-varying fading caused by multi-path and mobility is used to model the channel, along with random noise, burst interferences from other systems, and internal interferences among users when sharing spectrum resources.

The complex uncertainty of multimedia sources are often reflected by their non-stationary randomness. In particular, in network environments, the time distribution of compressed multimedia sources is likely non-stationary, whose asymptotic separability cannot be guaranteed. It is difficult to define the typical sequences in the length of $2^{nH(Q)}$ bits, and the lower bound of the acceptable error probability

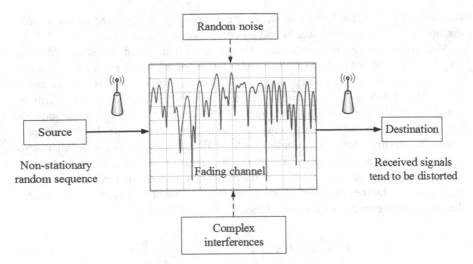

Fig. 1.3 An analytical model of wireless communication systems, where the transmission performance is mainly affected by both interferences and noise

may not be determined [5]. Besides, rate-distortion in wireless multimedia transmission has closed-form expressions only for a few of source models with specific distributions (e.g., Gaussian source, Bernoulli source, etc.).

Complex uncertainty in wireless channels are holistically affected by random noise, non-stationary fading, and complex interferences among multiple users. Fading caused by multi-path, scattering, reflection, and shadowing varies from time to time, and is likely a random process without ergodicity. Moreover, in mobile scenarios, Doppler shifts cause inter-carrier interference (ICI), delay spread causes inter-symbol interference (ISI), and multi-path propagation causes spatial coupling interference. Additionally, spectrum resources sharing among multiple nodes and multiple links brings about more complex spatial, time and frequency interferences, which are difficult to describe using simple statistical models [22, 23].

Complex uncertainty issues are almost everywhere in wireless communications. They are usually suggestive of a complex multiple-variable random process, for which a general and effective analytical model is often difficult to be built. The solution to this long-standing open problem would be a welcome breakthrough in wireless communication processing technology.

In conventional communications, the processing unit is *bit*, which matches the prerequisite, i.e., *i.d.d.*, of applying Shannon's theorems. Besides, for non-stationary but cyclic random processes, a concept of cyclostationary was proposed [24] to handle the random process with a relative invariance property per a cyclic granularity. However, a fixed granularity, either with bits or cycles, are suitable for ergodic process, but not for problems with complex uncertainty encountered in today's wireless communications.

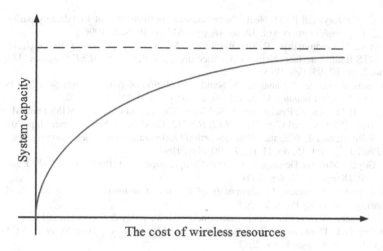

Fig. 1.4 Illustration of the "Marginal Effect". Although the budget of wireless resources can keep on increasing, the system capacity will meet its limit eventually

To be fair, conventional processing methods are very effective when processing simple data in simple systems subject to i.i.d., as indicated in Fig. 1.2. Nevertheless, with a continuous increase of complexity and uncertainty in services and systems, the deviation of sources and channels from i.i.d. distributions cannot be ignored, due to the significant observed decrease in the efficiency of such processing technology with i.i.d. models. In fact, a *marginal effect* has emerged in a variety of wideband services. As shown in Fig. 1.4, the increment of efficiency will be very marginal after a certain point although the cost of wireless resources continuously grows.

From our discussion, it is clear that a new breakthrough in the development of emerging wireless communications is urgently needed; and its realization hinges around the development of new processing technologies.

References

1. D. McQueen, "The momentum behind LTE adoption," *IEEE Commun. Mag.*, vol. 47, no. 2, pp. 44–45, Feb. 2009.
2. 3GPP, *3GPP specification: Requirements for further advancements for Evolved Universal Terrestrial Radio Access (E-UTRA) (LTE-Advanced)* [Online]. Avialable: http://www.3gpp.org/DynaReport/36913.htm [Accessed 9 Jul. 2014].
3. C. E. Shannon, "A Mathematical Theory of Communication," *Bell Syst. Tech. J.*, vol. 27, pp. 379–423, 623–656, Jul. Oct., 1948.
4. D. J. C. MacKay, *Information Theory, Inference, and Learning Algorithms*, Cambridge: Cambridge University Press, 2003.
5. T. M. Cover, and J. A. Thomas, *Elements of Information Theory, 2nd Edition*, Hoboken: Wiley-Interscience, 2012.

6. D. J. C. MacKay, and R. M. Neal, "Near Shannon performance of low density parity check codes," *Electronics Letters*, vol. 32, no. 18, pp. 1645–1646, Sept. 1996.

7. E. Berruto, M. Gudmundson, R. Menolascino, W. Mohr, and M. Pizarroso "Research Activities on UMTS Radio Interface, Network Architectures, and Planning," *IEEE Commun. Mag.*, vol. 36, no. 2, pp. 82–95, Feb. 1998.

8. D. Knisely, S. Kumar, S. Laha, and S. Nanda "Evolution of Wireless Data Services: IS-95 to cdma2000," *IEEE Commun. Mag.*, vol. 36, no. 10, pp. 140–149, Oct. 1998.

9. A. Yaver, P. Marsch, K. Pawlak, and F. S. Moya, "On the joint usage of MIMO and Multiflow in evolved HSPA networks," in *Proc. IEEE ICC'12*, Cape Town, South Africa, Jun. 2012.

10. G. J. Sullivan, and T. Wiegand "Rate-distortion Optimization for Video Compression," *IEEE Sig. Proc. Mag.*, vol. 15, no. 11, pp. 74–90, Nov. 1998.

11. V. K. Goyal, "Multiple Description Coding: Compression Meets the Network," *IEEE Sig. Proc. Mag.*, vol. 18, pp. 74–93, Sep. 2001.

12. D. Tse, and P. Viswanath *Fundamentals of Wireless Communication*, 1st ed., Cambridge: Cambridge University Press, 2005.

13. A. F. Molisch, *Wireless Communications*, 2nd ed., John Wiley & Sons Ltd., 2011.

14. W. R. Wu, G. L. Dong, and H. T. Li, *Engineering and Technology of Deep Space TT&C System*, 1st ed., Beijing: Science Press, 2013.

15. NASA, *Where are the Voyagers?*, Voyager 1, NASA, Jun. 2014. [Online]. Avialable: http:// voyager.jpl.nasa.gov/where/index.html. [Accessed 19 Jun. 2014].

16. D. D. Morabito, S. Shambayati, S. Finley, and D. Fort, "The Cassini May 2000 solar conjunction," *IEEE Trans. Antennas Propag.*, vol. 51, no. 2, pp. 201–219, Feb. 2003.

17. B. L. Hao, "The Description of Complexity and "Complexity Science"," *Sci.*, vol. 51, pp. 3–8, 1999.

18. R. Ahlswede, "Multi-way Communication Channels," in *Proc. of the 2nd International Symposium on Information Theory (ISIT)*, Tsahkadsor, USSR, pp. 23–52, Sept. 1971.

19. T. M. Cover, "Broadcast Channels," *IEEE Trans. Inf. Theory*, vol. 18, no. 1, pp. 2–14, Jan. 1972.

20. E. C. van der Meulen, "Three-Terminal Communication Channels," *Adv. Appl. Prob.*, vol. 3, no. 1, pp. 120–154, Spring 1971.

21. T. M. Cover, and A. A. El Gamal, "Capacity Theorems for the Relay Channel," *IEEE Trans. Inf. Theory*, vol. 25, no. 5, pp. 572–584, Sep. 1979.

22. P. Gupta, and P. R. Kumar, "The Capacity of Wireless Networks," *IEEE Trans. Inf. Theory*, vol. 46, no. 2, pp. 388–404, Mar. 2000.

23. M. Franceschetti, O. Dousse, D. N. C. Tse, and P. Thiran, "Closing the Gap in the Capacity of Wireless Networks Via Percolation Theory," *IEEE Trans. Inf. Theory*, vol. 53, no. 3, pp. 1009–1018, Mar. 2007.

24. W. R. Bennett, "Statistics of regenerative digital transmission," *Bell. Syst. Tech. J.*, vol. 37, pp. 1501–1542, Nov. 1958.

Chapter 2
Principle Shift: From Bit to Structure

2.1 On the Bit Representation

2.1.1 Representing Information with Bits

In the early twentieth century, as telegraph and telephone systems were being widely deployed in many countries around the world, researchers and scientists started to think about the speed and quality of message propagation. In 1924, H. Nyquist defined "intelligence" and proposed that the transmission speed of intelligence is proportional to the logarithm of the number of signal levels in a unit duration [1]. Let m denote the number of signal levels within each unit duration, the transmission speed of intelligence, W, can be expressed as

$$W = K \log m, \tag{2.1}$$

where K is a constant value. In addition, he also mentioned the possibility of the improvement in transmission rate using a certain "optimum" code.

In 1928, R. Hartley extended the concept of intelligence to "information" [2]. He introduced the "quantitative measure of information", and then concluded that the capacity is proportional to the bandwidth of the channel based on the observation with RLC circuits. Defining s as the number of symbols in each selection, the amount of information H can be written as

$$H = n \log s, \tag{2.2}$$

where n is the total number of selections. Besides, Hartley further confirmed the principle that the information is the outcome of a selection among a finite number of possibilities.

© The Author(s) 2015
J. Lu et al., *Structural Processing for Wireless Communications*, SpringerBriefs in Electrical and Computer Engineering, DOI 10.1007/978-3-319-15711-5_2

A general form of information representation is defined as entropy by Shannon in his notable work *A Mathematical Theory of Communication* in 1948 [3], where the entropy $H(X)$ is given by (1.2) in Chap. 1.

It is noted that, if the base of the logarithmic operation in (1.2) is 2, the unit of $H(X)$ is bit. That is to say, the information source X can be represented by $H(X)$ bits.

Furthermore, for continuous signal waveforms in wireless communications, they may be transformed to sequences of discrete samples by Nyquist-Shannon sampling theorem and then be represented in bits as well.

2.1.2 Channel Capacity by Bits

In fact, the bit representation aims not only to measure the uncertainty of information sources, but also to provide a measurement of capacity of communication channels.

Statistically, a communication channel is usually modeled as a triple consisting of an input alphabet, an output alphabet, and for each pair (i, o) of input and output elements a transition probability $p(i, o)$. Semantically, the transition probability is the probability that the symbol o is received given that i was transmitted over the channel.

In information theory, the channel capacity of a communication channel is the least upper bound on the data rate that can be reliably transmitted [4]. As early as in 1948, Shannon derived this upper bound for both discrete and continuous channels [3], which is called *Shannon capacity* nowadays. According to his results, capacity of a channel is given by the maximum of the average mutual information between the source and destination.

Now let us revisit the basic diagram of a wireless communication system as shown in Fig. 1.1 and consider a simplest channel model. Here, we use $\mathfrak{X}$ and $\mathfrak{Y}$ to denote discrete random variable spaces representing the input and output of the channel, respectively. $p(x|y)$ is defined as the conditional probability density function of x given y, which is determined by the channel. Let $p(y)$ be the marginal probability density of y, and $p(x, y)$ be the joint probability density, where

$$p(x, y) = p(x|y)p(y). \tag{2.3}$$

From [5], it is known that $H(X|Y)$ is defined as the conditional entropy of X over the joint XY ensemble, which can be expressed as

$$H(X|Y) = - \sum_{x \in \mathfrak{X}, y \in \mathfrak{Y}} p(x, y) \log p(x|y). \tag{2.4}$$

Since the entropy of X can be written as

$$H(X) = - \sum_{x \in \mathcal{X}} p(x) \log p(x), \qquad (2.5)$$

the average mutual information can be written as

$$\begin{aligned}
I(X;Y) &= H(X) - H(X|Y) \\
&= \sum_{x \in \mathcal{X}} \sum_{y \in \mathcal{Y}} p(x, y) \log \frac{p(x|y)}{p(x)} \\
&= \sum_{x \in \mathcal{X}} \sum_{y \in \mathcal{Y}} p(x, y) \log \frac{p(x, y)}{p(x)p(y)}.
\end{aligned} \qquad (2.6)$$

If variables X and Y are continuous, (2.6) can be rewritten as

$$I(X;Y) = \int_{\mathcal{X}} \int_{\mathcal{Y}} p(x, y) \log \frac{p(x, y)}{p(x)p(y)} dx dy. \qquad (2.7)$$

When the variables X and Y are mutually independent, the joint probability density can be written as $p(x, y) = p(x)p(y)$. As a result, (2.7) becomes zero. This implies that, if the channel quality is extremely bad, it is impossible to transmit any information from the source to destination. In other words, mutual information describes how much information can be shared between two communication nodes. The channel capacity thus may be defined as

$$C = \max_{p(x)} I(X;Y), \qquad (2.8)$$

where C and $I(X;Y)$ are all measured by bits.

2.1.3 The Limitation of Bit Representation

As discussed in Chap. 1, the conventional bit representation is confronted with challenges of complexity and uncertainty. To arrive at a fundamental understanding of the issues, we need to first revisit the general process of wireless communications in terms of space representation and mapping [6, 7].

As shown in Fig. 2.1, three spaces, consisting of information, coding, and signal spaces, are often involved in a communication system. Firstly, starting from the information space, a processing unit transforms the raw input data into representations suitable for transmission. Secondly, in coding space, a channel coding scheme transforms the information representation into a redundant form

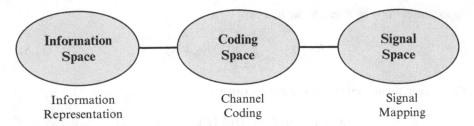

Fig. 2.1 Three spaces involved in wireless communications

capable of correcting errors caused by transmission. Finally, coded information is mapped to signal space with modulation, where new waveforms may be constructed through pre-coding to adapt to channel propagation conditions. In fact, information representation, channel coding and modulation are the three main aspects of the task of adapting services to wireless propagation environments undergoing dynamic changes.

It should be noted that the adoption of the bit representation in the information space suggests that the processing methods in the following steps are likely all based on bits. More importantly, the representation was originally proposed based on the assumption that the information source and the transmission channel are both independent stationary stochastic processes. Accordingly, errors occurring in the transmission process are also assumed independent. However, the burst of errors and the unequal relevance of source data will lead to the formation of complex uncertainty, which results in error propagation which may seriously degrade the quality of communications.

Besides, when channel coding is mainly designed based on the bit representation, the combination of bits is extremely huge even under moderate length of code-word. Design complexity and transmission performance may not be well balanced. While algebraic coding has low processing complexity, its performance is often limited, making it difficult to eliminate the influences of noises and interferences effectively.

It turns out that the uncertain errors and the complex processing scale that we have to face are often far beyond the processing capability of conventional communications with bit representations. This in turn provides us an impetus to develop a more reliable processing method to replace the bit representation in order to deal with issues of complexity and uncertainty.

2.2 Inspirations from Other Fields: A Structural Perspective

All along, engineers and technicians are very good at imitating nature to handle diverse complex design tasks. It has been noted that structures, of different kinds mimicked from or inspired by nature, often play important roles in transforming complex problems into those relatively easy to treat. For instance, in the history of development of the airplane, learning from birds helped the airplane pioneers to avoid solving from scratch many complex issues on aerodynamic analysis and design. Faced with the issues of complexity and uncertainty in wireless communication systems, some existing research results in other research areas may offer valuable inspirations.

2.2.1 Structure Design in Architecture

In architecture, structures are fundamental elements whose combinations form various elegant buildings. Figure 2.2 shows two representative examples of architecture structures, from ancient to modern.

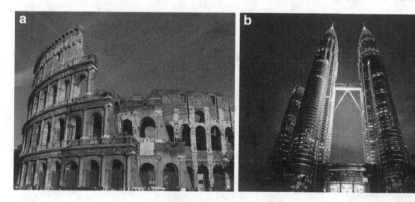

Fig. 2.2 Two examples of architecture structures: (**a**) the Colosseum from Ancient Rome; (**b**) the Petronas Twin Towers in Kuala Lumpur, Malaysia

In the history of architecture, it may be seen that bricks have been used for thousands of years as the main building material. In artifices ranging from the ancient Roman Colosseum shown in Fig. 2.2a to 19-century Victorian houses in Britain, bricks can always be found. Bricks, stacked up piece by piece, form the basic elements in many of these structures. However, when it comes to a large and complex buildings, construction by these "bricks" requires large labor force over a long period; worse yet, the whole body of such buildings may be unstable.

In modern architecture, steel-frame structural frameworks are vital indeed to the construction of large-spanned buildings and skyscrapers such as the Petronas Twin Towers in Kuala Lumpur shown in Fig. 2.2b. These frameworks usually consist of steel-framing with a few basic structures, and the design of the basic structures may change according to the target height of the skyscrapers. The building style where "bricks" were treated as basic elements has been abandoned, and the structural frameworks are instead regarded as whole entities when designing and constructing these skyscrapers. In this manner, large labor forces had been saved, while and more stable architecture at extreme heights sprung up to astound the world.

It is obvious that the design of structures is a major task in the construction of buildings. Is it possible to extend and generalize this structural perspective in architecture to other fields?

2.2.2 Structural Biology

Structural biology [8–10] is another excellent example where complex problems are solved effectively by adopting the structural perspective.

It is known that the real world is composed of various kinds of atoms, and one of the goals of biology is to analyze the inherent characters of organisms. However, due to the countless combinations of atoms, researches that regard atoms as basic units must be very complex and thus difficult.

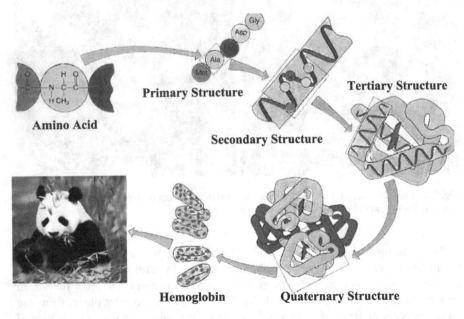

Fig. 2.3 An example for illustration of the basic principles of structural biology where the amino acids are the basic units for analyzing the properties of organisms

As an important milestone, structural biology provides us a new perspective by which many breakthroughs in the development of biology were made possible. In particular, the composition of organisms is based on specific elementary structures, namely, amino acids which consist of carbon, hydrogen, oxygen, nitrogen atoms with well-organized structures. The beauty is that, albeit the arbitrary combinations of atoms is nigh infinite, the number of types of amino acids we have discovered is quite limited, and the structures of amino acids are meaningful enough when used to explain the principle of the biosphere. As Fig. 2.3 shows, the basic structures of amino acids may further form hierarchical structures, including primary, secondary, tertiary, and quaternary structures. As the level of biological structures increases, they embody more complex parts and even macroscopic biological structures of living things such as various body cells, tissues, and organs, until forming the whole bodies of animals or plants.

Biological analysis based on specific finite structures will not only be working with much smaller numbers of possible instances, but will also exhibit meaningful rules [11, 12]. For example, if we know that some proteins are composed of certain amino acids, the biological function of those proteins will be correspondingly revealed to some extent. As we know, such methodology with structural perspective has already accomplished some major achievements in biology.

Can this structural perspective enlighten us when considering the design of structural information for communication systems? Intuitively, atoms in biology may regard as analogous to bits in communications, while the molecular structure of amino acids may be considered as analogous to hypothetical information structures, which we expect to discover or design to establish a new perspective of processing technology for wireless communication systems.

2.2.3 Cellular Structure

The structural perspective can already be found in the field of wireless communication networks. As is well known, supporting large numbers of concurrent mobile users over the radio is a very complex problem due to very limited spectrum and mutual interference among the users (as shown in Fig. 2.4a). Thanks to the introduction of cellular networks [13], this complex problem is quickly simplified with cellular structures.

As shown in Fig. 2.4b, the cellular structure divides randomly distributed users into hexagonal regions (cells), thus the number of users in each cell becomes limited, while frequency resource may also be reused among different cells. Exploiting the attenuation characteristics of radio communications, interference amongst cells may be reduced and that between users may be restricted within single cell. As such, the complexity of multi-user communications is decomposed by base stations which are only in charge of limited numbers of users in the corresponding cellular structure; in this manner, the processing scale for interference cancelation, etc., may be significantly reduced with the help of the cellular structure. The cellular structure has been proven to be effective in overcoming the problems of spectrum resource shortage and mobile handset power limitations.

The concept of cellular networks, as one of the top ten inventions that affect the communication area throughout, still dominates architecture design in communication networks. Its continued success serves as an inspiring proof for the effectiveness of a structural approach in response to the complexity and uncertainty of communication problems.

2.3 From Bit to Structure

As of now, is it possible to make the structural perspective applicable, and to herald a new era of prosperity in wireless communications research?

2.3.1 The Structure of IP Packets

In mobile Internet, the basic units transmitted between mobile nodes are IP packets rather than bits [14]. An IP packet complies with the principal Internet protocol for relaying data across network boundaries. Its routing function underlies every function of the Internet; it essentially establishes the Internet and keeps it together. In particular, IP has the task of delivering packets from the source host to the destination host solely based on the IP addresses in the packet headers.

As is shown in Fig. 2.5, an IP packet defines a packet structure that encapsulate the data to be delivered. It also defines addressing methods that are used to label the datagram with source and destination information. Although each IP packet is composed of bits, the function and importance of different bits in the packet may be different. Therefore, one may observe that IP groups some correlated bits and forms an intermediate layer structure, rather than treating bits independently.

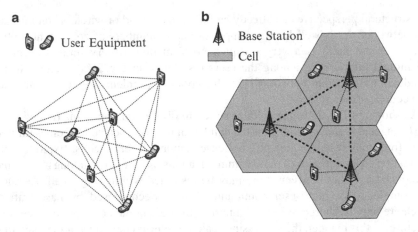

Fig. 2.4 Multi-user communications in wireless networks: (**a**) without cellular structure; (**b**) with cellular structure

Accordingly, the measurement of reliability in mobile Internet is packet loss rate rather than bit-error rate (BER). Hence, it is obvious that protecting the structure of IP packets is much more critical than protecting single bits. It will be seen from

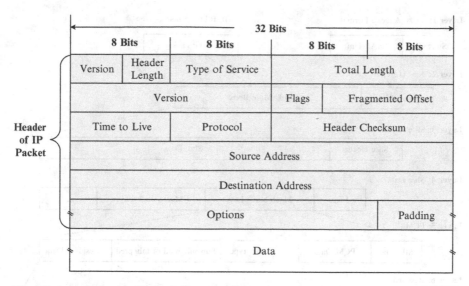

Fig. 2.5 A structure of an IP packet which consists of header and payload data

this case that bits are not necessarily the most effective processing units in practical information communications. Instead, some structures that are made up of bits may really be used to measure the reliability of communications. In other words, to better understand the essence and requirements of practical communication systems, it may be suggested that one should focus more on specific structures that capture the properties of wireless transmissions.

2.3.2 The Structure of Multimedia Streams with H.264

Similarly, transport of multimedia streaming data also relies on specific structures such as the example depicted in Fig. 2.6. In fact, the data structures for multimedia are carefully designed. For instance, the real time transport protocol (RTP) likely utilizes the structure of multiple layers which assume different responsibilities [15].

Due to the special header structure for multimedia streaming data, the importance of different bits varies based on their functions. If some bits in the structured header are lost or incorrect, the quality of source image or voice may suffer serious degradation. Hence, the structures of multimedia streams not only provide an effective way to represent data in a structured style, but also imply the significance of protecting the data structure rather than individual bits.

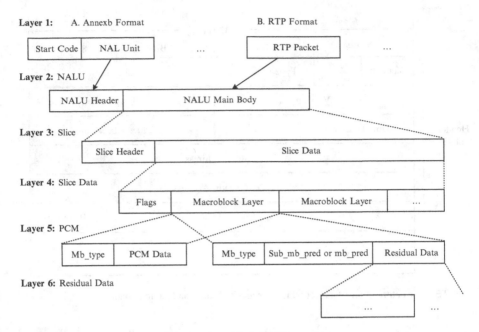

Fig. 2.6 A structure of multimedia streams with the H.264 standard

2.3.3 Additional Inspirations

In the classic Shannon information theory, *bit* is the minimum unit of information processing, and source or channel is simplified by the assumption of the asymptotic equipartition property (AEP)[3]. This assumption is based on a consideration of "statistical average" processing which encodes information as bits with equal status. As a result, the evaluation criteria of transmission performance is mostly BER.

However, most of transmitted information possess physical significance, especially in the case of multimedia information. Recent studies [16, 17] show that people are usually much more concerned with the subjective audio-visual experience, which often do not agree with the objective indicators of a communication system, such as BER. More importantly, processing methods one the bit level cannot effectively deal with the sources and channels with non-stationary and non-independent statistical characteristics due to the typical huge processing scales.

In analogy to the hierarchical concept in structural biology as shown in Fig. 2.7a, the amino acids on the "intermediate level" form an effective bridge between the copious number of combinations of and the characteristics of proteins. In like manner, one may also explore an "intermediate level" between the original bits and the information contents to form a similar hierarchical framework. Note that, the intermediate units, as shown in Fig. 2.7b, may be regarded as a processing structure in wireless communications. If appropriate intermediate units may be found for information processing, the corresponding processing scale may be greatly reduced.

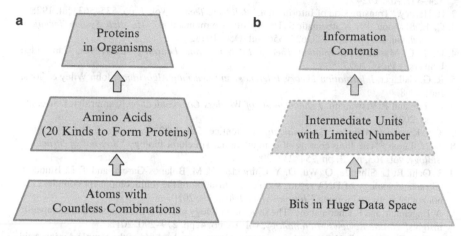

Fig. 2.7 An analogy between the hierarchical frameworks for structural biology and for information processing: (**a**) despite of the countless combinations of atoms in the natural world, only 20 kinds of amino acids are used to form the proteins in organisms, and limited combinations of amino acids may correspond to the functions of proteins; (**b**) to build a bridge between the bits in data space and the contents of information, intermediate units for information processing may also be necessary with significantly reduced processing scale

In summary, if a transition from bits to structures can be discovered and realized, the principles of contemporary communications will be shifted to the one where information structures form the basis of communication theories. However, the structures of IP packets and multimedia streams mentioned above are only simple concrete examples. More general theoretical problems need to be considered for a comprehensive study of structural processing technology.

In the following chapters, we will introduce some examples of structural processing technologies with promising and meaningful research results in wireless communications.

2.4 Conclusions

To address the issues of complexity and uncertainty in wireless communications, this chapter begins by reviewing the conventional bit representation of information and its associated problems. After clarifying the necessity of introducing the structural perspective for information processing, some existing approaches using structures are presented as references for stimulating a new framework of structural processing technology for future wireless communications.

References

1. H. Nyquist, "Certain Factors Affecting Telegraph Speed," *Bell Syst. Tech. J.*, vol. 3, pp. 324–352, Apr. 1924.
2. R. Hartley, "Transmission of Information," *Bell Syst. Tech. J.*, vol. 7, pp. 535–563, Jul. 1928.
3. C. E. Shannon, "A Mathematical Theory of Communication," *The Bell System Technical Journal*, vol. 27, pp. 379–423, 623–656, Jul. Oct., 1948.
4. D. J. C. MacKay, *Information Theory, Inference, and Learning Algorithms*, Cambridge University Press, 2005.
5. R. G. Gallager, *Information Theory, Inference, and Learning Algorithms*, John Wiley & Sons, Inc., 1968.
6. D. Tse, and P. Viswanath, *Fundamentals of Wireless Communication*, Cambridge University Press, 2005.
7. T. S. Rappaport, *Wireless Communications*, Prentice Hall, 2002.
8. A. N. Lupas, "The long coming of computational structural biology," *Journal of Structural Biology*, vol. 163, no. 3, pp. 254–257, 2008.
9. T. Ochi, B. L. Sibanda, Q. Wu, D. Y. Chirgadze, V. M. Bolanos-Garcia, and T. L. Blundell, "Structural biology of DNA repair: spatial organisation of the multicomponent complexes of nonhomologous end joining," *Journal of nucleic acids*, 2010.
10. A. Engelman, and P. Cherepanov, "The structural biology of HIV1: mechanistic and therapeutic insights," *Nature Reviews Microbiology*, vol. 10, no. 4, pp. 279–290, 2012.
11. A. R. Dalby, and A. F. Y. Poon, "A Comparative Proteomic Analysis of the Simple Amino Acid Repeat Distributions in Plasmodia Reveals Lineage Specific Amino Acid Selection," *PLOS One*, vol. 4, no. 7, 2009.
12. L. Y. Yampolsky, and M. A. Bouzinier, "Evolutionary patterns of amino acid substitutions in 12 Drosophila genomes," *BMC Genomics,* vol. 11, no. Suppl 4, pp. S10-12, 2010.
13. T. Farley, "Cellular Telephone Basics," Jan. 2006. [Online] Available: http://www.privateline.com/Cellbasics/Cellbasics02.html.
14. V. G. Cerf, and R. E. Kahn, "A Protocol for Packet Network Intercommunication," *IEEE Transactions on Communications*, vol. 22, no. 5, pp. 637–648, May 1974.
15. D. Marpe, "The H.264/MPEG4 advanced video coding standard and its applications," in *IEEE Communications Magazine*, vol. 44, no. 8, pp. 134–143, Aug. 2006.
16. J. You, U. Reiter, M. M. Hannuksela, M. Gabbouj, and A. Perkis, "Perceptual-based quality assessment for audio-visual services: A survey," *Signal Processing-image Communication*, vol. 25, no. 7, pp. 482–501, 2010.
17. B. Kunka, and B. Kostek, "An new method of audio-visual correlation analysis," *International Multiconference on Computer Science and Information Technology*, pp. 497–502, 2009.

Chapter 3
Processing Based on Information Structure

3.1 Structure Representation and Processing Method

3.1.1 The Design of Information Structure for Transmission

With reference to Fig. 2.7, "intermediate units" are desired such that the quality of communications can be closely controlled. The core of the problem of finding these units is to design a structure reflecting the inherent structural property of information to be transmitted. This kind of structure should be associated not only with service quality, but also with bit error distribution in transmissions. Nevertheless, multimedia data stream, for instance, usually has complex data structures which cannot largely be altered. The designed information structure has to fit the information features as well as to adapt to the channel characteristics. Moreover, the error rate of information structures may be calculated through the distribution of bit errors. As a result, it is possible to convert a non-i.i.d. problem into an i.i.d. like one.

As discussed in the preceding chapters, there exist logical structures in data communications. However, these structures may not suit harsh mobile environments well. In fact, their processing methods are likely based on the manipulation of bits; thus the structures may not be well protected, resulting in serious error propagation. It turns out that a proper design of structure is quite desirable. As one of the key design factors, the scale of structures must be taken into consideration, as it should not be too large or too small. If it is too large, the impact of channel impairments will be severe. However, if it is too small, the required processing complexity may be challenging. To strengthen the structural property, it is necessary to design such a universal and optimal data structure by considering the correlation of bit-streams, effectively improving the transmission efficiency and reliability as well as mitigating the problems with complexity and uncertainty in wireless communications.

© The Author(s) 2015
J. Lu et al., *Structural Processing for Wireless Communications*, SpringerBriefs
in Electrical and Computer Engineering, DOI 10.1007/978-3-319-15711-5_3

3.1.2   A Definition of MDU in Transmission

Note that, in conventional digital communications, a data unit is just a single bit or a single symbol; thus system performance is often measured by the use of bit-error rate (BER) or symbol-error rate (SER). However, in many cases, overall performance is eventually measured using word-error rate (WER) such as in speech communication [1] and in data communication [2]. If a transmission channel is memoryless and binary symmetric so that errors occur independently, there is a unique equivalence between BER/SER and WER. Such equivalence, on the other hand, does not exist if the transmission channel is subject to burst errors. As mentioned above, for highly compressed multimedia data, even a single bit error may corrupt a bundle of data such as in image transmission [3, 4] and in video transmission [5, 6]. In such case, BER or SER does not correlate well with the overall system performance. Instead, we need a more general form of WER which can serve as an excellent measure of the system performance.

To better represent such correlation, we introduce the work in [7], where an minimum data unit (MDU) is defined with the following features:

1. An MDU contains a number of bits which are not separable. That is, any residual errors, no matter how many, inside an MDU may cause the entire MDU to be destroyed or re-transmitted;
2. Different MDUs are well-separated in the sense that errors do not propagate across different MDUs;
3. The error rate of an MDU, denoted by $E_r(MDU)$, is indicative of the system performance.

Some examples are given to illustrate how the above defined MDU represents correlated data well. Consider a low bit-rate speech coder which encodes a block of speech, also known as a frame, at a time. In such case, an MDU contains the data of a speech frame, i.e., a speech word [1], and $E_r(MDU)$=WER. For image/video transmission, an MDU may be the interval between two consecutive synchronization flags in a compressed bit stream. Then, $E_r(MDU)$ is indicative of picture-block loss rate, which is a critical parameter to the received visual quality [8]. Likewise, in the transmission of raw data, an encoded packet can be considered as an MDU, and $E_r(MDU)$ is thus the probability of retransmission, which determines transmission throughput given a re-transmission protocol. Thereby, in general, MDU may better represent correlated data, and $E_r(MDU)$ may be a better performance measure accordingly.

3.1.3   Performance Analysis Based on MDU

Computation of the Error Rate of Minimum Data Unit, $E_r(MDU)$

Without loss of generality, consider a practical situation of channel coding based on systematic BCH codes, which are specified as $BCH(L(B), L(B'), t)$, where

B denotes a block code, B' is the information field of B, and their lengths are represented by $L(B)$ and $L(B')$, respectively, t is the error-correcting capability. Throughout, a wireless channel is modeled as a slow fading channel [9]. Note that most of the applications are likely to be in slow fading environments. Furthermore, assuming that the fading effects are approximately equal over a code-word, then the instantaneous bit error rate at time τ, denoted by $P_e(\tau)$, is approximately the same over the code-word after demodulation, no matter what kind of demodulator is used. Consequently, let $P_r(\tau, B, i)$ denote the probability that i errors occur in B at time τ. Then, it can be expressed as [7]

$$P_r(\tau, B, i) = \binom{L(B)}{i} P_e(\tau)^i [1 - P_e(\tau)]^{L(B)-i}. \tag{3.1}$$

To calculate $E_r(MDU)$, specifically, one shall study the relationship between the error rate of block codes and that of MDUs. Define $L(MDU)$ as the length of MDU and let $E_r(B)$ denote the error probabilities of B. For the case of $L(B) < L(MDU)$, it follows that [7]

$$E_r(MDU) = 1 - (1 - E_r(B))^{\frac{L(MDU)}{L(B')}}. \tag{3.2}$$

For small $E_r(B)$, one has [7]

$$E_r(MDU) \approx \frac{L(MDU)}{L(B')} E_r(B). \tag{3.3}$$

Then, consider the case of $L(B) > L(MDU)$, which is as,

$$E_r(MDU) = \sum_{i=t+1}^{L(B)} E_r(MDU|_i) P_r(\tau, B, i)$$

$$\approx \sum_{i=t+1}^{L(MDU)} E_r(MDU|_i) P_r(\tau, B, i) \tag{3.4}$$

where $E_r(MDU|_i)$ is the error probability of MDU given that the block B contains i bit errors. The approximation is due to the fact that $P_r(\tau, B, i)$ becomes extremely small as i becomes very large. In order to compute $E_r(MDU|_i)$, the following theorem is needed.

Theorem 3.1. *Let B be a block code, MDU be a minimum data unit with $L(MDU) < L(B)$. Assuming that B contains i error bits which are uniformly located in B with $i \leq min(L(MDU), L(B) - L(MDU))$, then [7]*

$$E_r(MDU|_i) = \frac{i}{i + \xi - 1} \tag{3.5}$$

where $E_r(MDU|_i)$ denotes the error probability of MDU under the condition of B containing i error bits and $\xi = \frac{L(B)}{L(MDU)}$.

Consequently, by substituting (3.5) into (3.4), one has [7]

$$E_r(MDU) \approx \sum_{i=t+1}^{L(MDU)} \frac{i}{i + \xi - 1} P_r(\tau, B, i). \tag{3.6}$$

To clearly understand the effect of different block codes on $E_r(MDU)$ in this case, consider the following numerical examples. Assume that $L(B) = 255$, $\xi = 1, 2, 3, 4$ (In practice, ξ should not be very large) and that different codes are constructed with different values of t. Then, a comparison in terms of $E_r(MDU)$ is shown in Fig. 3.1 using two different values of instantaneous BER, $P_e(\tau)$. A close observation of this figure indicates that with different values of ξ, $E_r(MDU)$ is very close for the same block code. In particular, the larger the t is, the smaller the relative difference. Therefore, for $L(B) > L(MDU)$, in a noisy fading channel which usually requires a large value of t, an important approximation can be obtained as [7]

$$E_r(MDU) \approx E_r(B). \tag{3.7}$$

Transmission Reliability

Basically, transmission reliability, denoted by ζ, refers to the percentage of information transmitted without loss. Since $E_r(MDU)$ is the probability of an MDU being destroyed or re-transmitted, it follows that [7]

$$\zeta = 1 - E_r(MDU). \tag{3.8}$$

From the above discussion, it turns out that $E_r(MDU)$ is determined by a block-code channel coding design. Now, it is necessary to discuss the reliability of block codes over correlated fading channels. Here, the reliability of block codes refers to the percentage of successfully transmitted code-words. Specifically, a wireless channel based on the GSM system configuration is considered, where the transmission rate $R_t = 270.83 \, kbit/s$ and the carrier frequency, f_c, is 900 MHz [10]. In addition, assuming vehicle speeds v of 2 and 50 miles/h, which corresponds to normalized Doppler frequencies, denoted by $f_d/timesT$, of 2.0×10^{-5} and 5.0×10^{-4}, respectively, where f_d is the maximum Doppler frequency, and $1/T$ is the symbol rate which is 135 ksps for the Gaussian minimum shift keying (GMSK) being used. Note that, both cases refer to very slow (or, highly correlated) fading conditions. Furthermore, suppose that the fading is subject to Rayleigh distribution.

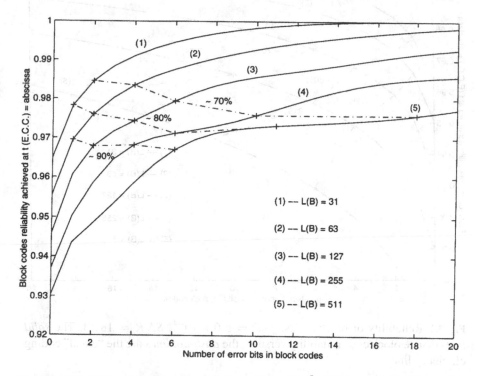

Fig. 3.1 Reliability of block codes. $f_d T = 2.0 \times 10^{-5}$, $SNR = 18$ dB. The *solid lines* denote block code reliability versus t; the *dash-dot lines* are the "equal" coding efficiency lines

Figures 3.1 and 3.2 show the block codes reliability of different BCH codes over the above specified channels with an signal-to-noise ratio (SNR) of 18 dB. In particular, note that the equal-efficiency lines along with coding efficiencies are plotted in the figures. It is shown that, in both slow fading channels, the reliability becomes worse as the block code size becomes larger at the same coding efficiency. This phenomenon becomes more explicit when the coding efficiency is higher. Likewise, Figs. 3.3 and 3.4 show the block codes reliability of different BCH codes over the specified mobile channels with an SNR of 9 dB and two-branch selection diversity. From these two figures, similar observation can be obtained. That is, longer codes tend to have lower reliability under the condition of the same coding efficiency.

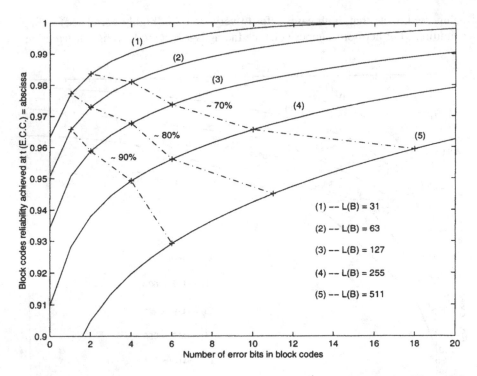

Fig. 3.2 Reliability of block codes. $f_d T = 5.0 \times 10^{-4}$, $SNR = 18$ dB. The *solid lines* denote block code reliability versus t; the *dash-dot lines* are the "equal" coding efficiency lines

Transmission Efficiency

The bit-expansion caused by channel coding overhead is one of the effect factor of overall transmission efficiency. As long as different channel coding schemes with different error-correcting capabilities achieves corresponding amount of bit-expansion, in a consideration of the simplicity problem, binary BCH codes are again used as the channel codes in [2]. BCH codes combine both reasonable burst error correction capability and dependable error detection. The ability of error detection is a significant factor for invoking re-transmission and/or post-processing. Moreover, the parameters of binary BCH code are given bellow [7]:

- Block length: $2^m - 1$;
- Number of parity-check digits: $\leq mt$;
- Minimum distance: $\geq 2t + 1$,

where $m(m \geq 3)$ is a positive integer and $t(t < 2^{m-1})$ is the error-correcting capability. When the value of t is low (otherwise, the detection will have very low

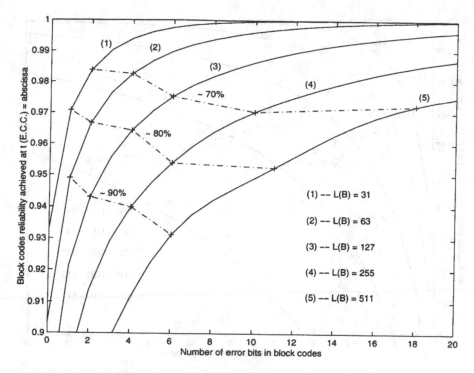

Fig. 3.3 Reliability of block codes. $f_d T = 2.0 \times 10^{-5}$, $SNR = 9\,\text{dB}$. Two-branch diversity is included. The *solid lines* denote block code reliability versus t; the *dash-dot lines* are the "equal" coding efficiency lines

coding efficiency), the number of parity-check digits are determined by mt [2]. β indicates the normalized bit-expansion that results from channel coding, this gives [7],

$$\beta \leq \frac{2^m - 1}{2^m - mt - 1} - 1 = \frac{mt}{2^m - mt - 1}. \tag{3.9}$$

Further, α is the normalized bit-expansion required for error-resilience enhancement in source coding, such as with re-synchronization insertion. Then, the overall normalized bit-expansion is given by [7]

$$(1 + \alpha) \times (1 + \beta) - 1 = \alpha + \beta + \alpha\beta \tag{3.10}$$

and the transmission efficiency is [7]

$$\eta \geq \frac{1}{(1 + \alpha) \times (1 + \beta)}. \tag{3.11}$$

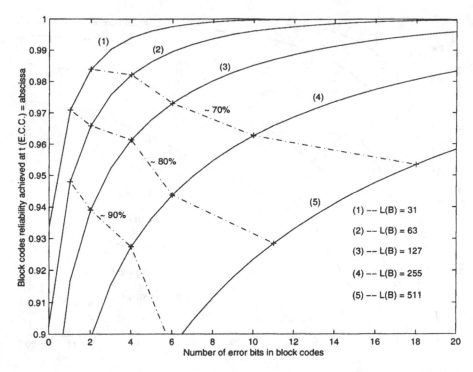

Fig. 3.4 Reliability of block codes. $f_d T = 5.0 \times 10^{-4}$, $SNR = 9\,\text{dB}$. Two-branch diversity is included. The *solid lines* denote block code reliability versus t; the *dash-dot lines* are the "equal" coding efficiency lines

In the equation above, it could be noticed that when the value of t is low, the equalities in (3.9) and thus in (3.11) are true.

Using the Eqs. (3.9)–(3.11), the scheme of designing a block code is discussed, like whether the given coding efficiency $E_r(MDU)$ should be minimized or maximized.

Consider an MDU with certain length. If $L(B) > L(MDU)$, then according to (3.7) it is better to have a small value of $L(B)$ if we want to achieve a small $E_r(MDU)$ for a given channel coding efficiency or to improve the coding efficiency for a given $E_r(MDU)$. For $L(B) < L(MDU)$, $L(B)$ tends to be larger. This can be best explained by an example: assuming that $L(MDU) = 255$ and channel coding efficiency is 80 %, then according to (3.3) and Fig. 3.2 (by referring to Figs. 3.1, 3.3, or 3.4),

$$E_r(MDU)|_{L(B)=63} \approx \frac{L(MDU)}{L(B')} E_r(B) = 5.0 \times 2.71\,\% = 13.55\,\%$$

$$E_r(MDU)|_{L(B)=127} \approx \frac{L(MDU)}{L(B')} E_r(B) = 2.58 \times 3.23\% = 8.33\%$$

$$E_r(MDU)|_{L(B)=255} \approx \frac{L(MDU)}{L(B')} E_r(B) = 1.23 \times 4.39\% = 5.40\%$$

In the conclusion of the above, the block code best contains one MDU where

$$L(B') = L(MDU) \tag{3.12}$$

to minimize $E_r(MDU)$ for a given coding efficiency or to maximize transmission efficiency under a certain $E_r(MDU)$. This result is found to be used for various correlated fading channel model with corresponding normalized Doppler spreads, SNR, and a limited number of diversity branches where $N \leq 3$.[1]

As long as the MDU have different values of length, different block codes may be found for revelent MDU lengths in terms of Eq. (3.12). The codes with minimal $E_r(MDU)$ or the maximal efficiency would be selected at the optimal ones.

The results are significantly useful for making a tradeoff between transmission efficiency η, and transmission reliability ζ. In this case, due to error-resilience enhancement, if a source coding algorithm using fixed length coding with no extra bit-expansion, η is only determined by the channel coding efficiency, and Eq. (3.12) provides a method for finding the optimal combined source-channel coding (CSCC). Meanwhile, if variable length coding is used, where an extra error-resilience capability is required, CSCC would be affected by the other factors.

3.2 Wireless Multimedia Transmission Based on the MDU Structure

3.2.1 The Effect of Errors on Multimedia Transmission

It is noted that multimedia data is usually compressed before transmission to save bandwidth, but being highly sensitive to channel errors. In the case of image transmission, for example, standards of compression, such as the JPEG, typically employ variable length coding (VLC) to achieve high-compression ratios. A characteristic of VLC is that a bundle of picture blocks (PBs) may be destroyed by the occurrence of even a single bit error. This is the well-known error propagation property of variable length coding. Furthermore, when the image data are transmitted over fading channels, such error propagation will become much more severe due to burst errors. As a result, lots of consecutive PBs may be corrupted at

---

[1] $N \geq 4$ is considered impractical in this study.

the same time. Figures 3.5 and 3.6 show a consecutively compressed bit-stream, and its corresponding original image, respectively. In fact, there exist structures inside the bit-stream, as indicated by the boundary lines. However, no clear intervals defined among them. As shown in Fig. 3.7, once errors occur, the boundaries may be shifted, eventually leading to error propagation and accumulation. It can be further observed from Fig. 3.8 that, burst channel errors corrupt almost entire rows of consecutive PBs.

In the case of video, there has been considerable interest in the transmission of H.264 coded video sequences over mobile network. This is mainly due to the excellent performance of H.264 in low-bit-rate video applications, which are very suitable for bandwidth-limited mobile networks. However, H.264 coded video data are also extremely sensitive to channel errors, especially to burst errors. Due to the use of predictive coding in H.264, channel errors may cause an entire frame or even consecutive frames to be lost, resulting in transmission interruptions.

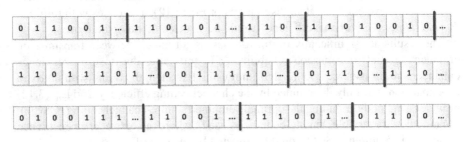

Fig. 3.5 A compressed image bit-stream

The complex effects of channel errors on multimedia transmission are yet to be comprehensively studied. Practical wireless communication often involves non-stationary and multi-variate random processes. As such, errors do not independently occur. For example, the errors caused by burst interference in space communications, and deep fading in mobile communications, often appear continuously with a strong correlation. Hence, the average BER may not be a good indicator of the quality of the communication. Moreover, the effects of errors are likely not isolated. Even a single bit error may cause a loss of data synchronization or error propagation, resulting in large segments of corrupted data, even leading to a communication interruption. It is worth noting that conventional communication processing is bit-based, with BER minimization as the design target; thus it is difficult to provide a remedy to this difficulty. In order to effectively achieve a reliable and efficient multimedia transmission under complex error environments, new processing methods must be explored.

Fig. 3.6 The original image corresponding to the bit-stream shown in Fig. 3.5

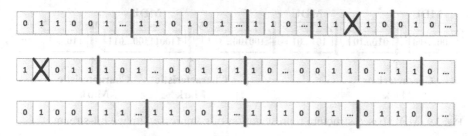

Fig. 3.7 The received bit-stream with bit errors

3.2.2 Structure Protection to Combat Error Propagation

As mentioned above, with the conventional bit-based information representation, the compressed bit-stream without clear defined intervals may lead to error propagation when bit errors occur. The utilization of structure representation may alleviate this problem. In Fig. 3.9, a consecutive compressed bit-stream is separated into units of grouped bits using characteristic markers, whereas the units correspond to PBs within an image, as shown in Fig. 3.10. Note that, the sizes of all PBs may be the same, but the length of corresponding units are likely different due to the usage of VLC. Each unit is an MDU, as defined in the previous section [7]. In Fig. 3.11, once bit errors occur in an MDU, error propagation is confined to the MDU, only affecting the corresponding PB of the image, as shown in Fig. 3.12.

Fig. 3.8 The corrupted image due to error propagation

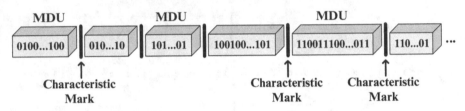

Fig. 3.9 MDU based structure representation of compressed image bit-stream

The idea of MDU-based representation is known as error isolation, which tries to restrict errors within limited areas. On the other hand, a difficult problem one has to face is finding a method to construct such a structure so as to adapt to both services' own data structure and the channel statistics.

In MPEG-4 (moving picture experts group phase 4) and H.264 standards, resynchronization marker and data partitioning methods are used for error isolation. These markers can be easily distinguished from all other code words, which are usually followed by header information. So long as the resynchronization marker is detected, the decoder can correctly decode the received information. However, the structure simply formed by the resynchronization markers may not be well suited to wireless channels.

In order to achieve better error isolation, data partitioning may be employed such that data between two synchronization markers can be further divided into smaller logical unit by inserting secondary synchronization codes. That is, for a given slice or a group of blocks (GOB), all macro-block headers, motion vectors, and discrete cosine transform (DCT) coefficients of macro blocks are placed in different

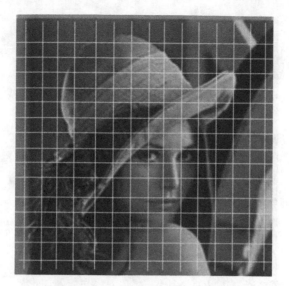

Fig. 3.10 A sample image segmented with PBs

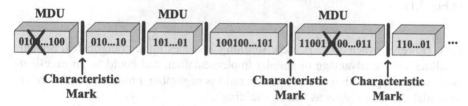

Fig. 3.11 The received MDU represented bit-stream with errors

logical units. If error occurs in the logical unit which contains DCT coefficients, the preceding logical units which contains header and motion information can still be decoded. With the use of resynchronization markers and data partitioning, different structures may be designed to fit wireless channels in order to improve the reliability and quality of transmission. Nevertheless, as analyzed in Sect. 3.1.3, proper design of MDU, or a structure, should also be able to effect a good tradeoff between transmission reliability and efficiency.

3.3   Block Shuffling as a Structure Processing Method for Image Transmission

The block shuffling technique intends to isolate erroneous image blocks even in correlated fading channels, in order to reduce the difficulty of error concealment at the decoder. Typically, block shuffling may be performed either in the spatial domain or in the transform domain. Wang et al. suggested that block shuffling should be performed in the transform domain to prevent degradation of the compression ratio

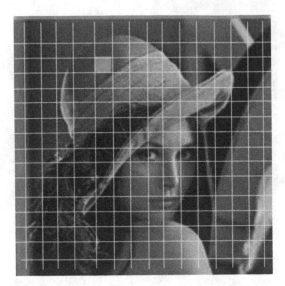

Fig. 3.12 The image with erroneous blocks corresponding to the erroneous MDUs in Fig. 3.11

[11]. However, when VLC is used, shuffling becomes more complicated. Spatial shuffling has the advantage of simpler implementation, and could be an excellent choice if the degradation in compression ratio is negligible. From now on, we refer to spatial shuffling simply as "block shuffling".

3.3.1   Block Shuffling with Random Re-ordering

The basic idea of block shuffling is to re-order the PBs within an image frame based on a specific pre-defined "shuffling pattern". This shuffling pattern could be designed such that neighboring PBs are separated as far as possible after shuffling. Once channel errors happen, consecutive shuffled PBs might be destroyed. After de-shuffling, these corrupted blocks are spread across the entire image. This is significant, since most error concealment methods inpaint the destroyed PBs by exploiting the spatial correlations among neighboring PBs. Therefore, if de-shuffled corrupt PBs are spread out and surrounded by good PBs, they can be concealed with much greater effectiveness. In this study, the shuffling pattern is created through adopting the algorithm shown below:

1. Generate a sequence of random numbers corresponding to all PBs inside the input image.
2. Scan the PBs in the image by the ascending order of these random numbers.

Note that the PB size used in block shuffling should be the same as that applied in the image codec. For example, if the JPEG codec employs 8×8 DCT, then a PB should contain 8×8 pixels.

As the JPEG baseline coding standard is generated according to PB, the DCT coefficients of a PB are assumed to be independent of those of other PBs. While block shuffling is operating, the transformed coefficients within each block are not altered. Moreover, the DCT coefficients of every PB are further categorized into the DC coefficient (the top-left one) and the AC coefficients (all other coefficients)[12]. Based on the JPEG standard, the DC and AC coefficients are conveyed individually. Particularly, zigzag scanning and run-length coding are the two main steps in the processing of the AC coefficients within every PB. Hence, block shuffling will not influence the compression ratio of the AC coefficients as the blocks are encoded regardless of their locations in the image. Whereas after quantization, the DC coefficient of every PB is extracted and subjected to differential pulse-code modulation (DPCM). While block shuffling is operating, the relative locations of the DC coefficients will change. Therefore, the redundancy between continuous DC coefficients might be reduced, decreasing the compression ratio of the DC coefficients.

3.3.2 Block Shuffling Versus Bit-Interleaving

Superficially, block-shuffling and bit-interleaving are both methods adopted to mitigate the impacts of burst errors on image transmission. However, they differ in three aspects, namely implementation, functionality, and performance.

There is extremely little extra hardware complexity owing to block shuffling. It can be simply obtained through reordering the interaction between video camera and the memory address of the image. Unless the image format is changed, the shuffling pattern does not need to be altered during image transmission. Thus, no additional transmission overhead is demanded for block shuffling. Furthermore, as block-shuffling operates before source encoding and de-shuffling after source decoding, no additional delay is involved.

On the contrary, bit-interleaving/de-interleaving normally occurs in the transmission phase (after channel coding and before channel decoding). Specifically, the I/O bandwidth of bit-interleaving is restricted by the transmission rate. In case of a low transmission rate, a large interleaving depth would necessitate excessive delay. Moreover, extra memory is demanded for interleaving and de-interleaving. Thus, block-shuffling is much more resource-efficient in image transmission.

Simulations are performed over the channels specified in Sect. 3.1.3. Image LENA with 256×256 pixels coded at 0.85 bit/pixel is used to test the proposed system, which is depicted in Fig. 3.13 from [7]. Moreover, the parameters of the slow Rayleigh fading channels are generated using Jakes' model [13].

Figures 3.14–3.17 show the simulation results of transmitting image LENA over the proposed system with vehicle speeds of 5 and 30 miles/h, which correspond

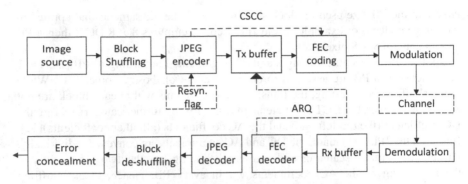

Fig. 3.13 A block diagram of the mobile image transmission system [7]

Fig. 3.14 Image LENA with 256×256 pixels at 0.85 bit/pixel through a mobile channel with $f_d T = 5 \times 10^{-5}$. $E_r(MDU) = 2.7\%$, $PSNR = 23.13$ dB

to $f_d T = 5 \times 10^{-5}$ and 3×10^{-4}, respectively. Let *Resyn* denotes the re-synchronization interval in terms of PBs. $BCH(255, 191, 8)$ and $Resyn = 4$, which implies a transmission efficiency as high as 70% with $E_r(MDU)$ being below 4%, are selected. From Figs. 3.14 and 3.16, it is observed that the $E_r(MDU)s$ in both fading channels are well controlled under a predefined value of 4%, thereby, confirming the validity of the design of information structure. By incorporating an error concealment algorithm, the reconstructed images in Figs. 3.15 and 3.17 present very good visual quality with peak signal-to-noise ratio (PSNR) of around 29 dB, verifying the usefulness of structure processing.

Fig. 3.15 Image LENA with 256×256 pixels at 0.85 bit/pixel through a mobile channel with $f_d T = 5 \times 10^{-5}$ using error concealment. $PSNR = 28.62\,\text{dB}$

Fig. 3.16 Image LENA with 256×256 pixels at 0.85 bit/pixel through a mobile channel with $f_d T = 3 \times 10^{-4}$. $E_r(MDU) = 2.4\%$, $PSNR = 23.89\,\text{dB}$

Fig. 3.17 Image LENA with 256×256 pixels at 0.85 bit/pixel through a mobile channel with $f_d T = 3 \times 10^{-4}$ using error concealment. $PSNR = 29.19\,\text{dB}$

3.4 Conclusions

In this chapter, MDU is defined as a universal structure representation of data, which is well suited to the wireless channels and the inherent source data structures. It is shown that structures could be protected to combat error propagation, where the effects of bit errors may be fully controlled within the MDUs. As an example, a generalized representation of correlated multimedia data is provided, and a combined source and channel coding scheme with simple BCH codes is designed to account for both the coded image characteristics and the wireless channel statistics, thereby achieving a good trade-off between transmission efficiency and information loss. Furthermore, block-shuffling and error-concealment techniques are included as essential means of structure processing to compensate for the information loss so as to guarantee acceptable visual quality of reconstructed images.

References

1. JTC(AIR), "Final Text for PACS Licensed Air Interface (TAG 3)," *J-STD 014*, Jun. 1995.
2. S. Lin, and D. J. Costello, "Error control coding: fundamentals and Applications," *Prentice Hall, Inc.*, 1983.
3. K. I. Chan, J. Lu, and J. C-I Chuang, "Block Shuffling and Adaptive Interleaving for Still Image Transmission over Rayleigh Fading Channels," *IEEE Trans. on Vehicular Technology*, vol. 48, no. 3, pp. 1002–1011, May 1999.

4. J. Lu, M. L. Liou, and K. Ben Letaief, "Efficient Image Transmission over Wireless Channels," *Proc. IEEE* 1997 *Inter. Symp. on Circuits and Systems, ISCAS'97*, Hong Kong, pp. 1097–1100, Jun. 1997.
5. H. Gharavi, and W. Y. Ng, "H.263 Compatibal Video Coding and Transmission," *Proc. Workshop on Wireless Image/Video Commun.*, Loughborough UK, pp. 115–120, Sep. 1996.
6. J. Lu, M. L. Liou, K. Ben Letaief, and J. C-I Chuang, "Error resilient transmission of H.263 coded video over mobile networks," *Proc. IEEE* 1998 *Inter. Symp. on Circuits and Systems, ISCAS'98*, Monterey, USA, Jun. 1998.
7. J. Lu, "Mobile Image/Video Transmission for Wireless Multimedia Communications," *Hong Kong University of Science and Technology*, May 1998.
8. J. Lu, M. L. Liou, K. Ben Letaief, and J. C-I Chuang, "Mobile image transmission using combined source and channel coding with low complexity concealment," *Signal Processing: Image Communication*, vol. 12, no. 2, Apr. 1998.
9. T. S. Rappaport, "Wireless communications, principles & practice," *Prentice-Hall*, 1996.
10. M. Mouly, and M.-B. Pautet, "The GSM system for Mobile Communications," *Europe Media Duplication*, S.A., 1993.
11. Y. Wang, Q.-F. Zhu, and L. Shaw, "Maximally Smooth Image Recovery in Transform Coding," *IEEE Trans. on Commun.*, vol. 41, no. 10, pp. 1544–1551, Oct. 1993.
12. G. K. Wallace, "The JPEG Still Picture Compression Standard," *Communications of the ACM*, vol. 34, no. 4, pp. 31–44, Apr. 1991.
13. W. C. Jakes, "Microwave mobile communications," *IEEE Press*, 1974.

Chapter 4
An LDPC Code Design with Sub-matrix Structure

4.1 Revisiting LDPC Code and Its Construction

Per the classical noisy-channel coding theorem [1], namely, *the random coding theorem*, an error correction code should be long enough with a well-selected generator matrix in order to achieve optimal performance. Accordingly, LDPC codes were invented by Robert Gallager in 1963 [2], and several types of construction methods were also proposed afterwards. Meanwhile, different decoding algorithms, including the famous belief propagation algorithm, were given. Nevertheless, due to very limited computational capability of computers in the 1960s, it was very difficult to evaluate and verify the performance of LDPC codes via computer simulations. In fact, LDPC codes were almost *"forgotten"* for quite a while after their birth. Until 1996, inspired by the invention of Turbo codes [3], and thanks to a significant progress in computer technology, LDPC codes were *"rediscovered"* [4] and attracted more and more researchers' interests thereafter.

Aided by the Gaussian approximation method, the performance of LDPC codes have been proven to approach the Shannon limit [5]. It is noted that various LDPC codes have already been adopted in several standards, such as DVB-S2 [6], CCSDS [7], and 802.11n/ac [8], etc. Different from codes which use simple coding polynomial, such as convolution code, RS code, and Turbo code, LDPC codes are a kind of linear block codes based on large-scale sparse random matrices, which are typically difficult to construct with both good performance and low complexity.

An LDPC coding matrix often contains thousands of rows and thousands of columns. As shown in Fig. 4.1, the elements of the matrix are represented by binary numbers, with bit "1" being sparsely and randomly located in each row and each column, resulting in a huge number of possible coding matrices to be evaluated. However, since the relationship between a sparse matrix and its coding performance has not yet been clearly revealed, there does not exist an established methodology for the optimal design of high performance LDPC codes. Instead, an LDPC code design usually relies on extensive computer search.

© The Author(s) 2015
J. Lu et al., *Structural Processing for Wireless Communications*, SpringerBriefs in Electrical and Computer Engineering, DOI 10.1007/978-3-319-15711-5_4

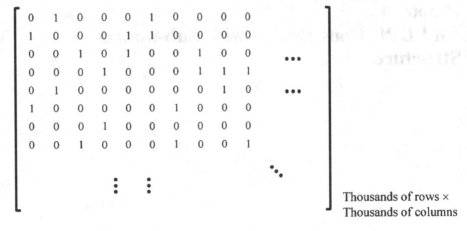

$$\begin{bmatrix}
0 & 1 & 0 & 0 & 0 & 1 & 0 & 0 & 0 & 0 \\
1 & 0 & 0 & 0 & 1 & 1 & 0 & 0 & 0 & 0 \\
0 & 0 & 1 & 0 & 1 & 0 & 0 & 1 & 0 & 0 \\
0 & 0 & 0 & 1 & 0 & 0 & 0 & 1 & 1 & 1 \\
0 & 1 & 0 & 0 & 0 & 0 & 0 & 0 & 1 & 0 \\
1 & 0 & 0 & 0 & 0 & 0 & 1 & 0 & 0 & 0 \\
0 & 0 & 0 & 1 & 0 & 0 & 0 & 0 & 0 & 0 \\
0 & 0 & 1 & 0 & 0 & 0 & 1 & 0 & 0 & 1 \\
\end{bmatrix}$$

Thousands of rows ×
Thousands of columns

Fig. 4.1 An LDPC code matrix

Theoretically speaking, in order to find an LDPC code matrix with optimal performance, complete traversal through all the possible distributions of bit "1" elements in the matrix is required; but the construction time of such a code would be unbearable. For example, a common LDPC code with a block length of 2,000, a code rate of 1/2, and an average degree distribution of (3, 6), may require a huge number of searches as given by,

$$(C_{1,000}^3)^{2,000} \approx 10^{16,000} \tag{4.1}$$

Even using the partition method with cyclic identity sub-matrix[1] size 125, the number is still high. That is,

$$(C_{1,000/125}^3)^{2,000/(2\times125)} \approx 10^{13} \tag{4.2}$$

Assuming that we use a powerful computer to evaluate the performance of the LDPC code matrices with a simulation speed at 1 min per matrix, the overall time for construction and evaluation will be about 18 million years. In fact, this bit-based construction method makes an optimal LDPC code search a nearly impossible task.

What's more, the implementation complexity for a bit-based construction is often too high to be applicable in practice due to a big resource consuming including storage and computation. Kim et al. [9] proposed a completely random constructor, and the constructed code-word is only 0.04 dB apart from the Shannon limit which is shown in Fig. 4.2. In this figure, the code length is 10^7 in bits, the column weight is 200, the maximum number of iterations is 2,000, and the average number of iterations is of 800–1,100. Considering its encoding or decoding complexity based

---

[1]Without loss of generality, we also classify cyclic identity sub-matrix as a bit-based method.

on the above parameters, both may exceed the computational capability of existing devices, and the above LDPC code-word with good performance is de facto difficult to be practical in reality.

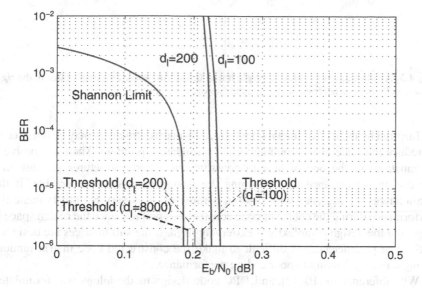

Fig. 4.2 The performance of Kim's LDPC code-word which is 0.04 dB apart from Shannon limit [9], and d_l denotes the maximum column weight of the code matrix

As a result, it is hard to effect a good trade-off between performance and complexity using the bit-based construction method.

4.2 Structured LDPC Code Design

Optimized LDPC code matrix design is typically a large scale combinatorial optimization problem with millions of variables. Conventional bit-based construction methods for LDPC codes often focus on low-level individual elements with reference to Fig. 2.7; but the coding performance is very difficult to control. In fact, the relationship between the distribution of bit-elements and coding performance is very complex and yet to be discovered. The question then is that, is it possible to design an intermediate element, a structure instead of bit, to balance the performance and complexity? Figure 4.3 shows an illustrative example of a simple fractal transform, which uses a series of binary trees, structures, as the intermediate level to compose a complex tree.

Fig. 4.3 An illustrative example of a simple fractal transform, where the complex tree of the *right* most may be constructed from the *left* structures progressively

Inspired by the idea of fractal transform, sub-matrix structures may be introduced to reduce the scale of the combinatorial optimization problem. Then, the problem is translated to the one of how to construct sub-matrix structures guaranteeing the coding performance while keeping the design complexity acceptable. If the sub-matrices are designed with random parameters, it is very hard to evaluate the performance during the course of code construction process since the search space is huge and the design complexity is extremely high. If the sub-matrices are designed with fixed parameters, it is difficult to make the constructed code matrix random enough, usually leading to poor coding performance.

With reference to [10–12], an LDPC code design methodology was formulated towards an excellent tradeoff between complexity and performance. As shown in Fig. 4.4, the right part of the coding matrix is a pre-defined structure that may achieve a low complexity, while the left part is a pseudo-random structure constructed gradually from sub-matrices designed with variable parameters in Galois field (GF), GF sub-matrices for simplicity, in order to obtain good coding performance. This way, the performance evaluation is more efficient when the sub-matrices based construction is combined with the pre-defined structure. Besides, the girth constraint may be well met by the design freedom of the GF sub-matrices. It is noted that the freedom of a GF sub-matrix of two dimensional structure may be about 100 times more than the one dimensional structure of a cyclic identity matrix [13–16], likely guaranteeing coding performance with sufficient randomness of a code matrix.

The design procedure may be described in detail as follows in Fig. 4.5. First, a pseudo-random sub-matrix is designed with two parameters in GF, determining bit locations within a permutation matrix. Using this sub-matrix as the basic structure, different sub-matrices may be composed with altering parameters in GF. Second, a base matrix is designed such that a complete LDPC code matrix may be constructed through a progressive expansion with the constructed GF sub-matrices. It is well verified that, LDPC codes that consist of GF sub-matrices may have excellent features of high coding performance and low complexity. This method based on GF sub-matrices, or, structures, may evolve an LDPC coding construction

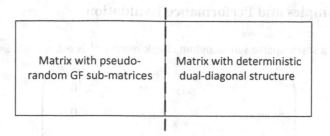

Fig. 4.4 An illustrative diagram of structured LDPC code design: *left part* with GF sub-matrices for good performance, and *right* with a regular structure for low complexity

from a generally extensive search to a structured design with significantly reduced complexity.

As a comparison, an LDPC code with the same parameters of the one in Sect. 4.1 is redesigned with GF sub-matrices. Since the entire matrix is symmetrical, the search space by the use of the structured method may be reduced to

$$(\frac{1}{1,000/125}C_{1,000/125}^3)^{2,000/(2\times125)} \approx 10^6 \tag{4.3}$$

Because of the freedom extension of GF sub-matrix structures, the girth constraint may be used to filter out those combinations of low coding performance. For the number of candidate matrices is reduced greatly with the structured design [17], a traverse search and corresponding performance evaluation may be conducted with an ordinary personal computer. As such, an LDPC code of good performance and low complexity may be obtained. An example LDPC code has been well designed and successfully applied to the mission of Chang E II lunar exploration.

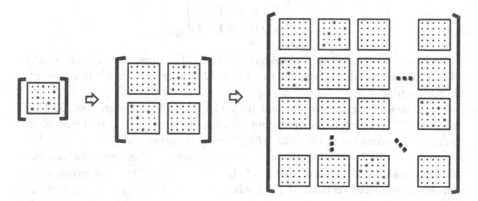

Fig. 4.5 An illustration of GF sub-matrices extension (left part of Fig. 4.4) with a high dimensional and nonlinear constraint

4.3 Examples and Performance Evaluation

First of all, a super-sparse semi-random check matrix H is established as.

$$
\begin{pmatrix}
 & \vdots & E_{127\times127} & 0 & 0 & 0 \\
A_{1,016\times1,016} & \vdots & 0 & E_{127\times127} & 0 & 0 \\
 & \vdots & 0 & 0 & \ddots & 0 \\
 & \vdots & 0 & 0 & 0 & E_{127\times127}
\end{pmatrix},
\tag{4.4}
$$

where

$$
E_{127\times127} =
\begin{bmatrix}
1 & 0 & 0 & \cdots & 0 \\
1 & 1 & 0 & \cdots & 0 \\
0 & 1 & 1 & \cdots & 0 \\
\vdots & \ddots & \ddots & \ddots & \vdots \\
0 & \cdots & 0 & 1 & 1
\end{bmatrix}_{127\times127}
,
\tag{4.5}
$$

And $A_{1,016\times1,016}$ comes from an extension of the basis matrix A_b, with an expansion coefficient L of 127. Basis matrix A_b can be expressed as

$$
A_b =
\begin{pmatrix}
1 & 1 & 0 & 0 & 1 & 0 & 1 & 1 \\
0 & 1 & 1 & 0 & 1 & 1 & 1 & 0 \\
0 & 0 & 1 & 1 & 1 & 1 & 0 & 1 \\
0 & 0 & 0 & 1 & 1 & 1 & 1 & 1 \\
1 & 0 & 0 & 0 & 1 & 1 & 1 & 1 \\
0 & 0 & 1 & 0 & 1 & 1 & 1 & 1 \\
0 & 1 & 0 & 0 & 1 & 1 & 1 & 1 \\
1 & 0 & 0 & 1 & 0 & 1 & 1 & 1
\end{pmatrix}_{8\times8}
.
\tag{4.6}
$$

Each zero element of A_b is extended to a 127×127 dimension all-zero matrix, while each one element is extended to a 127×127 dimensions matrix designed as a GF sub-matrix.

Let α denote a primitive element of a Galois field represented as $GF(2^m)$, and $L = 2^m - 1$ denote the spreading coefficient, where m is a prime number.[2] Then, all the elements in the field $GF(2^m)$ may be expressed as $0 = \alpha^\infty$, $1 = \alpha^0, \alpha^1, \alpha^2, \ldots, \alpha^{L-1}$. In addition, since m is the prime number, from the theorems of Galois field, the sequence $\alpha^i \cdot (\alpha^j)^0, \alpha^i \cdot (\alpha^j)^1, \ldots, \alpha^i \cdot (\alpha^j)^{L-1}$ comprises all the non-zero elements of the field $GF(2^m)$, where $0 < i < L, 0 \le j < L$ and both i, j are integers. If irreducible polynomial $f(x)$ with power m is a primitive polynomial in $GF(2)$, let $f(\alpha) = 0$, then $GF(2^m)$ can be constructed accordingly.

---

[2]With some constraints, m can be a positive integer.

Table 4.1 GF sub-matrix parameters of an example LDPC code

Row \ Column	0	1	2	3	4	5	6	7
0	111,3	114,39	\	\	69,23	\	11,50	33,77
1	\	35,112	7,103	\	57,12	14,89	10,58	\
2	\	\	20,67	24,134	91,33	7,49	\	127,11
3	\	\	\	33,47	9,133	57,28	26,3	5,48
4	13,91	\	\	\	68,10	39,95	13,45	121,92
5	\	\	35,125	\	38,95	54,44	12,79	4,46
6	\	51,101	\	\	103,52	24,115	46,88	56,67
7	103,69	\	\	93,107	\	18,123	17,50	104,85

The corresponding value of the field element sequence $\alpha^i \cdot (\alpha^j)^0, \alpha^i \cdot (\alpha^j)^1, \ldots, \alpha^i \cdot (\alpha^j)^{L-1}$ may be written as $f(\alpha^i \cdot (\alpha^j)^0), f(\alpha^i \cdot (\alpha^j)^1), \ldots, f(\alpha^i \cdot (\alpha^j)^{L-1})$, which is a pseudo-random permutation of the positive integer sequence $1, 2, \ldots, L$. Denote the pseudo-random permutation as $(f(\alpha^i), f(\alpha^j))$, where $f(\alpha^i)$ is the shifting factor, and $f(\alpha^j)$ is the interleaving factor. By using the pseudo-random permutation sequence $f(\alpha^i \cdot (\alpha^j)^0), f(\alpha^i \cdot (\alpha^j)^1), \ldots, f(\alpha^i \cdot (\alpha^j)^{L-1})$, an extended matrix may be obtained with $L \times L$ dimension, where $L - 1$ and $f(\alpha^i \cdot (\alpha^j)^{L-1})$ are the row and column numbers of non-zero elements, respectively.

As a result, once the shifting and interleaving factors are given, a pseudo-random matrix of $L \times L$ dimensions may be obtained for a non-zero element of A_b. For zero element of A_b, they can be directly extended to an $L \times L$ dimensions all-zero matrix. That is, an LDPC code matrix A may be systematically designed by its base matrix A_b as well as the corresponding shifting and interleaving factors for individual non-zero elements of A_b.

In Table 4.1, the detailed parameters of the above example LDPC code are given. The first element is the shifting factor, and the second element is the interleaving factor. Slash lines imply those shifting and interleaving factors unavailable, corresponding to the zero elements of the base matrix A_b.

The code constructed with parameters in Table 4.1 is further evaluated with simulations. In particular, binary phase-shift keying (BPSK) is used to modulate the LDPC code with additive white Gaussian noise (AWGN), and belief prorogation algorithm is adopted to decode the LDPC code with maximum number of iterations of 32. It is shown in Fig. 4.6 that the constructed LDPC code with structured sub-matrices has excellent performance.

Furthermore, in order to accommodate multi-purpose applications, multi-length, multi-rate LDPC codes may be constructed based on structured GF sub-matrices (Table 4.2), including 8 code rates and 40 different codes as an example. Specifically, there are five sizes of sub-matrices, namely, 31×31, 63×63, 127×127, 255×255, and 511×511, in the code group. Each sub-matrix may correspond two different block lengths, and each code rate includes five different block lengths.

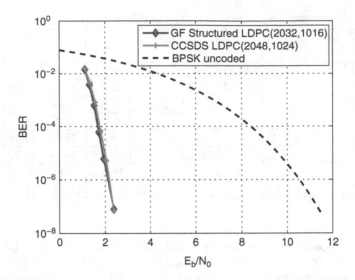

Fig. 4.6 Performance of a structured LDPC code with GF sub-matrices, with similar coding gain but much less construction complexity, as compared with the one defined in the CCSDS standard

Table 4.2 Multi-length multi-rate LDPC codes based on GF sub-matrices

Sub-matrix size	Code length							
	$\frac{1}{4}$	$\frac{1}{3}$	$\frac{1}{2}$	$\frac{2}{3}$	$\frac{3}{4}$	$\frac{4}{5}$	$\frac{5}{6}$	$\frac{7}{8}$
31	992	930	992	930	992	930	930	992
63	2,016	1,890	2,016	1,890	2,016	1,890	1,890	2,016
127	4,064	3,810	4,064	3,810	4,064	3,810	3,810	4,064
255	8,160	7,650	8,160	7,650	8,160	7,650	7,650	8,160
511	16,352	15,330	16,352	15,330	16,352	15,330	15,330	16,352

Performance of the above mentioned code group in Table 4.2 is depicted in Fig. 4.7. In simulations, BPSK modulation is employed and an AWGN channel is assumed. For simplicity, the values of energy per bit to noise power spectral density ratio, E_b/N_0, corresponding a bit-error rate (BER) of 10^6 of each code with a code length of about 16K, 4K, or 2K are given. For comparison, the theoretical performance of BPSK is also given as the dashed line.

4.4   Conclusions

In this chapter, the basic method of structured design for LDPC coding is introduced. Compared with conventional bit-based LDPC code construction, which usually relies on a heavy search, the proposed structured LDPC code with GF sub-matrices

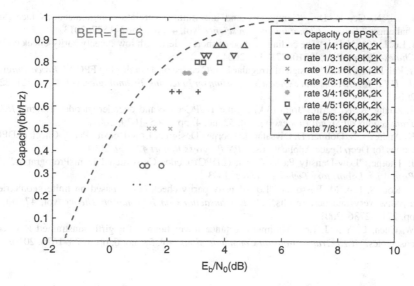

Fig. 4.7 Performance of structured multi-length and multi-rate LDPC codes with GF matrices

may achieve much better tradeoff between coding performance and complexity. It is noted that, the structured LDPC coding method takes advantage of both random coding and algebraic coding, resulting in near-optimal performance in terms of BER with low construction complexity.

References

1. C. E. Shannon, "A mathematical theory of communication," *Bell Syst. Tech J.*, vol. 27, pp. 379–423, 623–656, 1948.
2. R. G. Gallager, "Low Density Parity Check Codes," *IEEE Transactions on Information Theory*, vol. 8, pp. 21–28, 1962.
3. C. Berrou, A. Glavieux, "Near optimum error correcting coding and decoding: turbo-codes," *IEEE Transactions on Communications*, vol. 44, no. 10, pp. 1261–1271, 1996.
4. D. MacKay, R. Neal, "Near Shannon limit performance of low density parity check codes," *Electronics Letters*, vol. 32, no. 18, pp. 1645, 1996.
5. C. Sae-Young, T. J. Richardson, R. L. Urbanke, "Analysis of sum-product decoding of lowdensity parity-check codes using a Gaussian approximation," *IEEE Transactions on Information Theory*, vol. 47, no. 2, pp. 657–670, 2001.
6. European Standard, "Digital Video Broadcasting (DVB);Second generation framing structure, channel coding and modulation systems for Broadcasting,Interactive Services, News Gathering and other broadband satellite applications (DVB-S2)," *ETSI EN 302 307 V1.2.1*, 2009.
7. CCSDS Standard, "Low Density Parity Check Codes for Use in Near-Earth and Deep Space Applications," *Experimental Specification CCSDS 131.1-O-2*, 2007.
8. IEEE WIFI standard, "Wireless LAN media access control and physical layer specifications".
9. C. Sae-Young, J. Forney, T. J. Richardson, et al., "On the design of low-density parity-check codes within 0.0045 dB of the Shannon limit,", *IEEE Communications Letters*, vol. 5, no. 2, pp. 58–60, 2001.

10. J. Lu, N. Ge, B. Du, "Complexity and uncertainty: wireless communications face dual challenges," *Science China: infomation science,* vol. 43, pp. 1563–1577, 2013.
11. J. Lu, L. Yin, Y. Pei, "The coding method of short code-length low-density parity-check code," *Chinese Patent,* ZL 200910077183.6, 2004.
12. Y. Pei, L. Yin, J. Lu, "Design of irregular LDPC codec on a single chip FPGA," *Proceedings of the IEEE 6th Circuits and Systems Symposium on Emerging Technologies,* vol. 1, pp. 221–224, 2004.
13. T. Zhang, K. K. Parhi, "Joint (3,k)-regular LDPC code and decoder/encoder design," *IEEE Transactions on Signal Processing,* vol. 52, no. 4, pp. 1065–1079, 2004.
14. K. Andrews, S. Dolinar, D. Divsalar, J. Thorpe "Design of Low-Density Parity-Check (LDPC) Codes for Deep Space Applications," *IPN Progress Report 42-159,* 2004.
15. J. Thorpe, "Low-Density Parity-Check (LDPC) Codes Constructed from Protographs," *Jet Propulsion Laboratory Technical report,* 2003.
16. Y. Kou, S. Lin, M. Fossorier, "Low-density parity-check codes based on finite geometries: a rediscovery and new results," *IEEE Transactions on Information Theory,* vol. 47, no. 7, pp. 2711–2736, 2001.
17. W. Chen, L. Yin, J. Lu, "Minimum distance lower bounds for girth-constrained RA code ensembles," *IEEE Transactions on Communications,* vol. 58, no. 6, pp. 1623–1626, 2010.

Chapter 5
Pre-coding Design with Constellation Structures

5.1 Pre-coding in MIMO Systems

Equipping multiple antennas on both transmitter and receiver, MIMO may dramatically improve system performance. In particular, with spatial diversity, multiplexing, and interference mitigation, MIMO embodies a great breakthrough for wireless communication technology. As a result, MIMO has been widely adopted in various standards, e.g., IEEE 802.11, IEEE 802.16.

Nevertheless, in MIMO systems, the existence of space correlation and interference necessitates high processing complexity in the receiver. Via preprocessing at the transmitter based on full or partial channel state information (CSI), pre-coding may effectively reduce the interference among multiple streams or multi-users. Hence, research on pre-coding design algorithms and related techniques has attracted great attention in the regime of MIMO systems.

With a geometrical characterization, pre-coding refers to placing points in $O(N_t)$-dimensional space in such a way to optimize the corresponding criteria, as is shown in Fig. 5.1, where N_t is the number of transmit antennas. There may be three different possible limitations on the output of the pre-coder:

1. All pre-coding outputs are required to have exactly the same power P. It is required to choose points lying on the surface of a sphere of radius $\sqrt{N_t P}$.
2. All pre-coding outputs have power P or less. In this case, all points are required to lie interior to or on the surface of a sphere of radius $\sqrt{N_t P}$.
3. The average power of all pre-coding outputs is P or less. Obviously, individual output may have a greater squared distance than $N_t P$ but the average of the set of squared distances cannot exceed $N_t P$.

The first two conditions are simpler and the third condition is somewhat more general. In this brief, we derive the structured pre-coding algorithm based on these conditions.

© The Author(s) 2015
J. Lu et al., *Structural Processing for Wireless Communications*, SpringerBriefs
in Electrical and Computer Engineering, DOI 10.1007/978-3-319-15711-5_5

Fig. 5.1 Illustration of the
pre-coding design in the
geometrical view. The sphere
represents the
$O(N_t)$-dimensional space
with given power constraint.
The radius of the sphere is
$\sqrt{N_t P}$. The points in the
sphere represent the probable
output of the pre-coder.
The distance between any two
points denotes the Euclid
distance

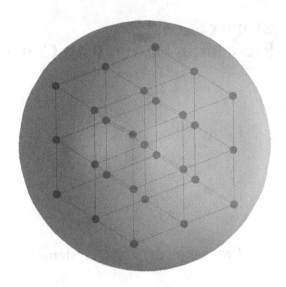

Actually, similar geometrical representation of mapping has been used since Shannon [1, 2]. In [2], a code-word of length n is thought of geometrically as a point in a n-dimensional Euclidean space. An optimal decoding system for a code is one which minimizes the probability of error, P_e, for the code. With the geometrical approach, finding the good codes is equivalent to placing points in the n space in such a way to minimize P_e. Exploiting the fact that the minimal probability of error, namely $P_{e,opt}$, is a function of a quotient A (the root square of the signal power divided by the noise power) by change of scale in the geometrical picture, the upper and lower bounds on $P_{e,opt}$ are obtained. These bounds are reasonably close together over an important range of values and give good estimates of $P_{e,opt}$.

From the above description, one can see clearly that the pre-coding design and [2] share a similar basic idea, i.e., a vector of points should be placed in a way to optimize some given criteria; the job is to design the placement method. For example, if the criteria is to maximize the achievable rate, the principle should be to place as many points in the sphere as possible while guaranteeing given probability of error.

Similarly, if the criteria is to minimize the probability of error, then the principle should be to make the distance between the points as large as possible with a given number of points. However, one may notice that, while they share similar basic idea, a fundamental difference lies with the channel matrix **H** of a MIMO system, which causes distortion to the transmitted signal, such that the sphere may be turned into an ellipsoid in the multi-dimensional space. Hence, during the pre-coding design for a MIMO system, the placement of the points needs to match the channel matrix while giving consideration to an additive noise.

In MIMO systems, the pre-coding algorithm can be divided into two categories: linear and nonlinear. Linear pre-coding algorithms include zero-forcing, minimum mean squared error (MMSE), Block Diagonalization, etc. Typical representation of a nonlinear pre-coding algorithm is the dirty paper coding (DPC). Due to the extremely high implementation complexity of DPC, a series of sub-optimal algorithms such as Tomlinson-Harashima, Channel Inversion, and Regularized Inversion are proposed. In comparison with many nonlinear pre-coding techniques, linear pre-coding is a good choice considering both capacity performance and implementation complexity [3]. Hence, we mainly consider the linear pre-coder in this brief.

As to the classification based on CSI, two cases are usually considered. On the one hand, the transmitter may have full knowledge of the CSI. In this case, the transmitter can make full use of the CSI to design proper transmit signal and optimally mitigate the spatial interference. On the other hand, the transmitter may have only partial CSI. In this case, one of the most effective methods is the finite rate feedback pre-coding design. Based on the statistical information of the channel, a group of optimized pre-coding matrices form a codebook, which is known at both transmitter and receiver. The receiver selects the proper pre-coding matrix from the codebook based on the estimation of the channel realization and given performance metric. The index of the selected pre-coding matrix is then fed back to the transmitter. In this chapter, we focus on the linear pre-coding design with full CSI, and the extension to the case with partial CSI is discussed in the conclusion.

For the linear pre-coder in a MIMO system with N_t transmit antennas and N_r receive antennas, the output of the pre-coding is the transmit signal $\mathbf{x} = [x_1, \cdots, x_{N_t}]^T$. An illustration of a linear pre-coder is given in Fig. 5.2, where $\mathbf{P}$ denotes a linear pre-coding matrix. Passing through the channel matrix $\mathbf{H}$ and noise $\mathbf{n}$, the received signal is given by $\mathbf{y} = \mathbf{Hx} + \mathbf{n}$.

In order to achieve channel capacity, the mutual information

$$I = H\left(\mathbf{y}\right) - H\left(\mathbf{y}|\mathbf{x}\right) = H\left(\mathbf{y}\right) - H\left(\mathbf{n}\right) \tag{5.1}$$

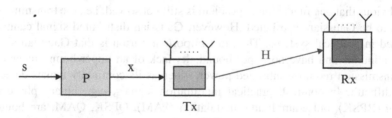

Fig. 5.2 Illustration of a linear pre-coder in a MIMO system

needs to be maximized. With the channel state information known at the transmitter, maximizing the mutual information is equivalent to maximizing the entropy of $\mathbf{y}$. The extension of Shannon information theory in multiple antenna systems [4, 5] shows that two conditions need to be satisfied to reach the capacity:

1. The transmit signal **x** is Gaussian distributed;
2. The covariance matrix of **x** satisfies the structure:

$$\Sigma_\mathbf{x} = \mathbf{V_H D^2 V_H^H} \tag{5.2}$$

where $\mathbf{V_H}$ is the right eigenmatrix of channel matrix **H**, and **D** is the diagonal matrix derived by the water-filling algorithm [6].

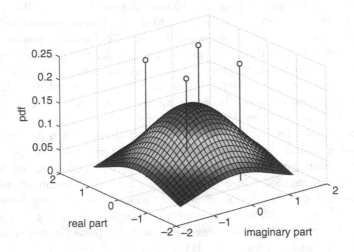

Fig. 5.3 Comparison of probability density function between Gaussian signal and QPSK signal. Clearly, the four black discrete points represent the QPSK constellation. And the colored continuous mesh represents the probability density function of Gaussian signal

While the discussion above shows that linear pre-coding may reach the capacity, it is obvious that one fundamental problem is still not solved, i.e., the transmit signal needs to be Gaussian distributed. However, Gaussian distributed signal cannot be realized in practical systems. The most important reason is that Gaussian signals are continuous and have no upper bound. Its lack of an upper bound means that its transmission requires unlimited power, whereas its continuity makes detection very difficult. In contrast, practical modulation signals, e.g., binary phase-shift keying (BPSK), pulse-amplitude modulation (PAM), QPSK, QAM, are bounded and discrete, ensuring their realizability with limited transmit power and easy detection. Figure 5.3 shows the probability density function comparison between the QPSK signal and the Gaussian signal. One can notice the tremendous difference between the two.

The obvious difference on the signal format brings great difference to the system performance. Figure 5.4 compares the spectral efficiency of QPSK signal and Gaussian signal respectively in a 2 × 2 MIMO Rayleigh channel. Two dashed lines are Gaussian input with and without pre-coding according to (5.2). And two solid lines are QPSK input with and without pre-coding. Comparing these two solid lines, one can see that employing the capacity achievable pre-coding design for Gaussian signal to practical modulation signal brings performance degradation instead of performance improvement.

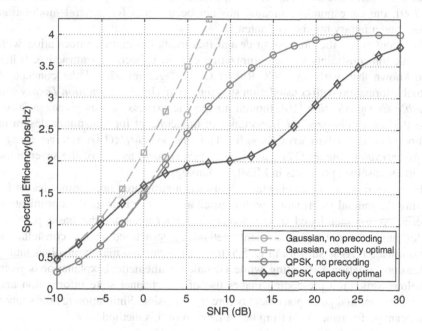

Fig. 5.4 Comparison of spectral efficiency between Gaussian signal and QPSK signal

This phenomena implies that it is of great importance to design effective pre-coding algorithm based on the constellation structure of practical modulation signals.

5.2 The Idea of Structured Pre-coding

Researchers have done some jobs in finding pre-coding design methods to tackle the problem shown above. One method is to maximize the diversity gain. This mainly involves maximizing the slope of the pairwise error probability but obviously cannot guarantee approaching the optimal performance. Another method aims at maximizing the minimum distance between constellations, i.e.,

$$d_{\min} = \min_{\substack{m \neq k \\ m,k=1,\cdots,M^{N_t}}} \left\| \mathbf{H}\mathbf{P}\left(\mathbf{x}_m - \mathbf{x}_k\right) \right\|^2 \tag{5.3}$$

where $\mathbf{x}_m$ and $\mathbf{x}_k$ denote the mth and kth probable constellation vector at the MIMO transmitter, respectively [7]. Since the minimum distance between constellation vectors has no closed form expression, finding the minimum distance pair needs exhaustive search and hence is very complicated. Although progress has been made in [7–9], the corresponding method has not been found for general modulation signals and arbitrary numbers of antenna.

In recent years, the pre-coding design that treats the mutual information with constellation constraints as the optimization criteria is becoming attractive. It has been known that, early in 1968, Robert G. Gallager introduced the concept of mutual information with constellation structures in his book *Information Theory and Reliable Communication* [10], though no specific expression was given. In 2005, Ezio Biglieri emphasized the importance of this concept for transmitter design in his book *Coding for Wireless Channels* [11]. In the book *MIMO Transceiver Design via Majorization Theory* [12], Paloar listed this concept as one of the seven most important unsolved problems in MIMO technology.

In the remainder of this chapter, we discuss a structured pre-coding design by treating the mutual information with constellation constraints as the optimization criteria. As we mentioned in the geometrical interpretation, the basic problem is designing the optimal placement method by structured signal constellation transformation. As is shown in Fig. 5.5, this transformation mainly involves multi-dimensional rotation and scaling, while a detailed mathematical explanation is given in below. Structured pre-coding makes use of the channel state information and designs the optimal point placement before transmission. Simulation results show a significant performance gain from the utilization of this method.

5.3 Pre-coding Optimization Algorithm with Constellation Structures

Consider a MIMO system with N_t transmit antennas and N_r receive antennas. $\mathbf{x} \in \mathbb{C}^{N_t}$ denotes the transmitted signal with zero mean and identity covariance. The received signal can be expressed as,

$$\mathbf{y} = \mathbf{HPx} + \mathbf{n} \tag{5.4}$$

where $\mathbf{H} \in \mathbb{C}^{N_r \times N_t}$ is the channel response matrix, $\mathbf{P} \in \mathbb{C}^{N_t \times N_t}$ the linear pre-coder matrix, and $\mathbf{n} \in \mathbb{C}^{N_r}$ the zero-mean circularly-symmetric Gaussian noise with covariance $\sigma^2 \mathbf{I}$, where $\mathbf{I}$ denotes the identity matrix. Moreover, the channel response matrix is assumed to be constant and known at both the transmitter and the receiver.

With finite-alphabet modulation, $\mathbf{x}$ is drawn from equiprobable constellation set with cardinality M. Then, the mutual information between $\mathbf{x}$ and $\mathbf{y}$ can be expressed as [13]

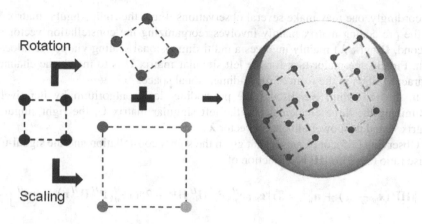

Fig. 5.5 Illustration of the idea of structured pre-coding design

$$I[\mathbf{H}, \mathbf{P}] = N_t \log M - \frac{1}{M^{N_t}} \sum_{m=1}^{M^{N_t}} \mathbf{E_n} \log \sum_{k=1}^{M^{N_t}} \exp(-d_{mk}) \tag{5.5}$$

where

$$d_{mk} = (\|\mathbf{HPe}_{mk} + \mathbf{n}\|^2 - \|\mathbf{n}\|^2)/\sigma^2 \tag{5.6}$$

and $\mathbf{e}_{mk} = \mathbf{x}_m - \mathbf{x}_k$. $\mathbf{x}_m$ and $\mathbf{x}_k$ contain N_t independent symbols from the M-ary signal constellation.

The objective is to design a linear pre-coder, $\mathbf{P}$, that maximizes the mutual information. Considering the conditions shown in the geometrical representation, the optimization constraint is given by $\mathrm{Tr}(\mathbf{PP}^H) \leq N_t$, which is equivalent to $\mathrm{Tr}(\mathbf{PP}^H) = N_t$ since $I[\mathbf{H}, \mathbf{P}]$ is an increasing function of the transmit power. Hence, the limit on the mutual information with finite-alphabet inputs is given by

$$I[\mathbf{H}] = \max_{\mathbf{P}:\mathrm{Tr}(\mathbf{PP}^H)=N_t} I[\mathbf{H}, \mathbf{P}] \tag{5.7}$$

As mentioned above, it is desirable to derive the optimal placing method with multidimensional rotation and scaling. In the view of mathematics, we denote the singular value decomposition (SVD) of the pre-coding matrix as $\mathbf{P} = \mathbf{U_P}\mathbf{Diag}(\sqrt{\lambda})\mathbf{V_P}^H$. For simplicity, let $\mathbf{U_P} = \mathbf{U}$, $\mathbf{V_P}^H = \mathbf{V}$. Clearly, $\mathbf{U}$ and $\mathbf{V}$ are unitary matrices and represent the multi-dimensional rotation of the precoder; $\mathbf{Diag}(\sqrt{\lambda})$ represents the scaling operation.

Similarly, the SVD of the channel matrix is given by $\mathbf{H} = \mathbf{U_H}\mathbf{Diag}(\sigma)\mathbf{V_H^H}$. Substituting the SVD decomposition of the pre-coding matrix and the channel matrix into (5.4) yields

$$\mathbf{y} = \mathbf{U_H}\mathbf{Diag}(\sigma)\mathbf{V_H^H}\mathbf{U}\mathbf{Diag}(\sqrt{\lambda})\mathbf{V}\mathbf{x} + \mathbf{n} \tag{5.8}$$

Accordingly, one may make several observations. First, the right singular matrix $\mathbf{V}$ of the pre-coding matrix mainly involves reorganizing the constellation vector $\mathbf{x}$. Second, $\mathbf{Diag}(\sqrt{\lambda})$ mainly involves a multi-dimensional scaling via power allocation. Finally, the major task for the left singular matrix $\mathbf{U}$ is to match the channel characterization in the form of multi-dimensional space.

In the following, we discuss the pre-coding design algorithm by iteratively optimizing the three elements, i.e., the left singular matrix $\mathbf{U}$, the right singular matrix $\mathbf{V}$ and the power allocation vector λ.

Observing (5.5), one can see that given the signal constellation and the signal-to-noise ratio (SNR), $I[\mathbf{H}]$ is a function of

$$\|\mathbf{HP}(\mathbf{x}_m - \mathbf{x}_k) + \mathbf{n}\|^2 = \mathrm{Tr}\left[\mathbf{e}_{mk}\mathbf{e}_{mk}^H\mathbf{P}^H\mathbf{H}^H\mathbf{HP} + 2\Re\left(\mathbf{e}_{mk}^H\mathbf{P}^H\mathbf{H}^H\mathbf{n}\right) + \mathbf{nn}^H\right]$$

where $\Re$ denotes the real part of a complex number. Obviously I changes based on the distribution of $\|\mathbf{HP}(\mathbf{x}_m - \mathbf{x}_k) + \mathbf{n}\|^2$ depending on $\mathbf{P}$ through $\mathbf{P}^H\mathbf{H}^H\mathbf{HP}$. As is shown in [14], setting the left singular vectors of $\mathbf{P}$ equal to the right singular vectors of $\mathbf{H}$ maximizes the mutual information for general channel conditions and arbitrary inputs with a given matrix $\mathbf{P}^H\mathbf{H}^H\mathbf{HP}$ and a specific power constraint. Hence, the optimal left singular vectors $\mathbf{U} = \mathbf{V_H}$ is adopted. As a result, the signal model can be simplified to

$$\mathbf{y} = \mathbf{Diag}(\sigma)\mathbf{Diag}(\sqrt{\lambda})\mathbf{Vx} + \mathbf{n}. \tag{5.9}$$

Given $\mathbf{U}$, the optimization problem is addressed over λ:

$$\begin{aligned} & \text{maximize } I(\lambda) \\ & \text{subject to } \mathrm{Tr}\left(\mathbf{PP}^H\right) = \mathbf{1}^T\lambda \leq N_t \\ & \qquad\qquad \lambda \succeq \mathbf{0} \end{aligned} \tag{5.10}$$

where $\mathbf{1}$ and $\mathbf{0}$ denote the column vector with all entries being one and zero, respectively.

In order to make the inequality constraints implicit in the objective function, the problem formulation may be rewritten as,

$$\text{minimize } f(\lambda) = -I(\lambda) + \sum_{i=1}^{N_t}\phi(-\lambda_i) + \phi(\mathbf{1}^T\lambda - N_t) \tag{5.11}$$

where

$$\phi(u) = \begin{cases} -(1/t)\ln(-u), & u < 0 \\ +\infty, & u \geq 0 \end{cases}$$

$t > 0$ sets the accuracy of the approximation [15].

Algorithm 1 Maximize mutual information over the power allocation Vector [16]:

1. Given a feasible vector λ, $t := t^{(0)} > 0$, $\alpha > 1$, tolerance $\epsilon > 0$.
2. Compute the gradient of f at λ, $\nabla_\lambda f(\lambda)$, as (5.12) and the descent direction $\Delta\lambda = -\nabla_\lambda f(\lambda)$.
3. Evaluate $\|\Delta\lambda\|^2$. If it is sufficiently small, then go to Step 6; else go to Step 4.
4. Choose step size γ so that $f(\lambda + \gamma\Delta\lambda) < f(\lambda)$ by backtracking line search.
5. Set $\lambda := \lambda + \gamma\Delta\lambda$. Go to Step 2.
6. Stop if $1/t < \epsilon$, else $t := \alpha t$, and go to step 2.

According to Proposition 1 in [16], the gradient of the objective function (5.11) is given by

$$\nabla_\lambda f(\lambda) = -\mathbf{R} \cdot \mathbf{vec}\left(\mathbf{Diag}^2(\sigma)\mathbf{VEV}^H\right) - \frac{1}{t}\left(\mathbf{q} - \frac{1}{N_t - \mathbf{1}^T\lambda}\right)$$

where $\mathbf{E} \triangleq \mathbb{E}\left\{[\mathbf{x} - \mathbb{E}(\mathbf{x}|\mathbf{y})][\mathbf{x} - \mathbb{E}(\mathbf{x}|\mathbf{y})]^H\right\}$ is the MMSE matrix, $\mathbb{E}(\mathbf{x}|\mathbf{y})$ denotes the conditional expectation of $\mathbf{x}$ given $\mathbf{y}$, and $\mathbf{R} \in \mathbb{R}^{N_t \times N_t^2}$ is a reduction matrix with entries given by $[\mathbf{R}]_{i,N_t(j-1)+k} = \delta_{ijk}$. If $i = j = k$, $\delta_{ijk} = 1$. Otherwise, $\delta_{ijk} = 0$. $q_i = 1/\lambda_i$ is the i-th element of vector $\mathbf{q}$. Thus, the steepest descent direction is chosen as

$$\Delta\lambda = -\nabla_\lambda f(\lambda).$$

Combining this search direction with the backtracking line search conditions [15], Algorithm 1 for the optimal power allocation vector, is developed. Convergence is ensured due to the concavity.

Finally comes the maximization of the mutual information over the right singular vectors $\mathbf{V}$ for a given λ, i.e.,

$$\begin{aligned}\text{maximize } & I(\mathbf{V}) \\ \text{subject to } & \mathbf{V}^H\mathbf{V} = \mathbf{V}\,\mathbf{V}^H = \mathbf{I}\end{aligned} \qquad (5.12)$$

It can be formulated as an unconstrained optimization in a constrained search space:

$$\text{minimize } g(\mathbf{V})$$

with domain restricted to the Stiefel manifold $\text{St}(n)$ [17]

$$\text{dom } g = \{\mathbf{V} \in \text{St}(n)\}$$

and

$$\text{St}(n) = \{\mathbf{V} \in \mathbb{C}^{n \times n} | \mathbf{V}^H\mathbf{V} = \mathbf{I}\}$$

Algorithm 2 Maximize the mutual information on complex Stiefel manifold [16]:

1. Given a feasible $\mathbf{V} \in \mathbb{C}^{n \times n}$ such that $\mathbf{V}^H \mathbf{V} = \mathbf{I}$.
2. Compute the descent direction $\Delta\mathbf{V}$ as (5.13). Set the step size $\gamma := 1$.
3. Evaluate $\|\Delta\mathbf{V}\|^2 = \text{Tr}\{(\Delta\mathbf{V})^H \Delta\mathbf{V}\}$. If it is sufficiently small, then stop; else go to Step 4.
4. Choose step size γ so that $g(\pi(\mathbf{V} + \gamma\Delta\mathbf{V})) < g(\mathbf{V})$ by backtracking line search.
5. Set $\mathbf{V} := \pi(\mathbf{V} + \gamma\Delta\mathbf{V})$. Go to Step 2.

Algorithm 3 Two-step algorithm to maximize the mutual information for a generalized pre-coder [16]:

1. *Initialization.* Set the left singular vectors of the pre-coder $\mathbf{U} := \mathbf{V_H}$. Specify a feasible λ and $\mathbf{V}$.
2. *Update power allocation vector*: Run Algorithm 1 given $\mathbf{V}$.
3. *Update right singular vectors*: Run Algorithm 2 given the obtained λ in Step 2.
4. Repeat Step 2 and Step 3 until convergence.

---

where the function $g(\mathbf{V})$ is defined as $-I(\mathbf{V})$. For each point $\mathbf{V} \in \text{St}(n)$, the search direction is given by [18]

$$\Delta\mathbf{V} = -\nabla_{\mathbf{V}} g(\mathbf{V}) = \nabla_{\mathbf{V}} I(\mathbf{V}) - \mathbf{V}(\nabla_{\mathbf{V}} I(\mathbf{V}))^H \mathbf{V} \tag{5.13}$$

where $\nabla_{\mathbf{V}} I(\mathbf{V})$ is the gradient of mutual information with respect to $\mathbf{V}$, given by $\mathbf{Diag}^2(\sigma)\mathbf{Diag}(\lambda)\mathbf{VE}$.

For an arbitrary matrix $\mathbf{W} \in \mathbb{C}^{n \times n}$, its projection $\pi(\mathbf{W})$ on the Stiefel manifold is defined as the point closest to $\mathbf{W}$ in the Euclidean norm

$$\pi(\mathbf{W}) = \arg\min_{\mathbf{Q} \in \text{St}(n)} \|\mathbf{W} - \mathbf{Q}\|^2$$

If the SVD of $\mathbf{W}$ is $\mathbf{W} = \mathbf{U_W}\boldsymbol{\Sigma}\mathbf{V_W}$, the projection can be expressed by $\mathbf{U_W}\mathbf{V_W}$ [19, Sec. 7.4.8].

Combining the search direction and the projection with the backtracking line search condition, Algorithm 2 which maximizes the mutual information over the right singular vectors $\mathbf{V}$ is given.

By combining Algorithms 1 and 2, the complete two-step approach is developed in Algorithms 3.

In the low SNR region, the algorithm may converge to the global optimum. Whereas for medium to high SNR, the method may converge to a local maximum theoretically, thus yielding near optimal performance.

5.4 Numerical Results

Figures 5.6 and 5.7 show the mutual information of the structured pre-coding algorithm versus SNR for BPSK and QPSK inputs, respectively. The performance is compared with other schemes such as those with maximum diversity in [20], maximum coding gain in [20, 21], and maximum capacity assuming Gaussian inputs in [22, 23].

From Figs. 5.6 and 5.7, one may see that the mutual information can achieve the upper bounded 1 and 2 bps/Hz, respectively, for BPSK and QPSK with high SNR. Although the maximum coding gain method [20] performs better than the maximum diversity method and the transmission without pre-coding, it is valid only for several kinds of antenna number and modulation type. In comparison, the structured design method may be used for an arbitrary antenna number and modulation type. Likewise, exploiting the degrees of freedom in the optimal left singular vectors, the optimal power allocation vector, and the local optimal right singular vectors simultaneously, the structured pre-coding algorithm may provide significant gains of mutual information in a wide range of SNR. For example, with input BPSK and 3/4 channel coding rate, the performance is about 4, 5.5, and 6 dB better than those of the maximum coding gain, the one without pre-coding, and maximum diversity methods, respectively. Moreover, it is observed that the structured pre-coding method achieves mutual information very close to maximum capacity with Gaussian inputs when the channel coding rate is below 0.6 for both BPSK and QPSK. This also outperforms the case of Gaussian inputs without pre-coding when the channel coding rate is below 0.9.

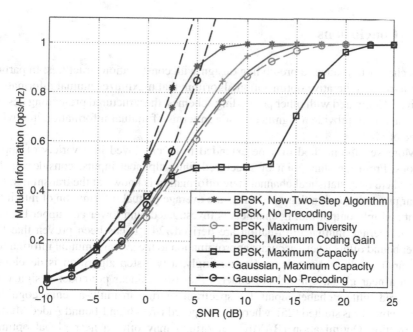

Fig. 5.6 Mutual information versus the SNR in the case of BPSK being employed [16]

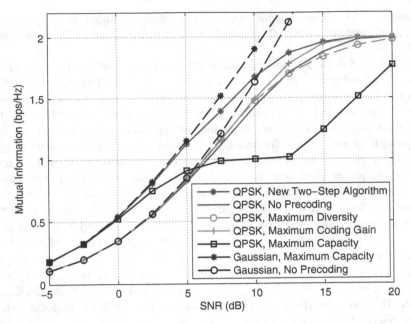

Fig. 5.7 Mutual information versus the SNR in the case of QPSK being employed [16]

5.5   Conclusions

This chapter introduces a pre-coding design with constellation structures. In particular, a two-step iterative optimization algorithm that maximizes mutual information is given. Compared with other pre-coding schemes, the structured pre-coding design algorithm may provide a significant gain in terms of mutual information in a wide range of SNR.

Moreover, the method may be extended to be employed in a variety of applications. From the standpoint of structured finite-alphabet inputs, considering the case where only statistical channel state information is known at the transmitter, the linear pre-coder design that maximizes the average mutual information of multiple-input multiple-output fading channels is investigated, and lower and upper bounds for the average mutual information are derived [24]. It has been proven that the lower bound offers a very accurate approximation to the average mutual information for various fading channels, and accordingly, a two-step algorithm is developed to get a near global optimal pre-coder. Likewise, a linear pre-coder design with structured finite-alphabet inputs for spectrum sharing in multi-antenna cognitive radio networks is studied [25], where the proposed Branch-and-bound Aided Mutual Information Optimization (BAMIO) algorithm may offer a near global optimal solution with only several iterations.

One may expect that the idea of structured pre-coding design will be applicable to various aspects of wireless communication scenarios, such as cooperative systems, virtual MIMO, etc.

References

1. C. E. Shannon, "Communicntion in the presence of noise," Proc I.R.E., 37, Jan. 1949, p. 10.
2. C. E. Shannon, "Probability of error for optimal codes in a Gaussian channel," *Bell System Technical Journal*, vol. 37, no. 3, pp. 611–656, May 1959.
3. F. Boccardi, F. Tosato, and G. Caire, "Precoding schemes for the MIMO-GBC," in *Proc. Int. Zurich Seminar on Commununications*, Zurich, Switzerland, pp. 10–13, Feb. 2006.
4. A. Goldsmith, S. Jafar, N. Jindal, and S. Vishwanath, "Capacity limits of MIMO channels," *IEEE J. Sel. Aeras Commun.*, vol. 21, no. 5, pp. 684–702, Jun. 2003.
5. G. Foschini, "Layered space-time architecture for wireless communication in a fading environment when using multi-element antennas," *Bell labs technical journal*, vol. 1, no. 2, pp. 41–59, 1996.
6. T. Cover and J. Thomas, *Elements of Information Theory*, 2nd ed. New York:Wiley, 2006.
7. L. Collin, O. Berder, P. Rostaing, et al., "Optimal minimum distance-based precoder for MIMO spatial multiplexing systems," *IEEE Trans. Signal Process.*, vol. 52, no. 3, pp. 617–627, Mar. 2004.
8. B. Vrigneau, J. Letessier, P. Rostaing, et al, "Extension of the MIMO precoder based on the minimum Euclidean distance: A cross-form matrix," *IEEE J. Sel. Signal Process.*, vol. 2, no. 2, pp. 135–146, Apr. 2008.
9. Q. Ngo, O. Berder, and P. Scalart, 'Minimum euclidean distance-based precoding for three-dimensional multiple input multiple output spatial multiplexing systems," *IEEE Trans. Wireless Commun.*, vol. 11, no. 7, pp. 2486–2495, July 2012.
10. R. G. Gallager, *Information theory and reliable communication.* 1968.
11. E. Biglieri, *Coding for Wireless Channels.* New York, NY: Springer, 2005.
12. D. Palomar, and Y. Jiang, *MIMO Transceiver Design via Majorization Theory.* Delft, The Netherlands: Now Publishers, 2006.
13. C. Xiao, Y. R. Zheng, and Z. Ding, "Globally optimal linear precoders for finite alphabet signals over complex vector Gaussian channels," *IEEE Trans. Signal Process.*, vol. 59, no. 7, pp. 3301–3314, Jul. 2011.
14. D. Palomar and Y. Jiang, *MIMO Transceiver Design via Majorization Theory.* Now Publishers Inc., 2006.
15. S. Boyd and L. Vandenberghe, *Convex Optimization.* Cambridge University Press, 2004.
16. W. Zeng, Y. R. Zheng, M. Wang, and J. Lu, "Linear precoding for relay networks: a perspective on finite-alphabet inputs," *IEEE Trans. Wireless Commun.*, vol. 11, no. 3, pp. 1146–1157, Mar. 2012.
17. P. Absil, R. Mahony, and R. Sepulchre, *Optimization Algorithms on Matrix Manifolds.* Princeton University Press, 2008.
18. J. H. Manton, "Optimization algorithms exploiting unitary constraints," *IEEE Trans. Signal Process.*, vol. 50, no. 3, pp. 635–650, Mar. 2002.
19. R. Horn and C. Johnson, *Matrix Analysis.* Cambridge University Press, 1985.
20. Y. Ding, J. Zhang, and K. Wong, "The amplify-and-forward half-duplex cooperative system: Pairwise error probability and precoder design," *IEEE Trans. Signal Process.*, vol. 55, no. 2, pp. 605–617, Feb. 2007.
21. Y. Ding, J. K. Zhang, and K. M. Wong, "Optimal precoder for amplify-and-forward half-duplex relay system," *IEEE Trans. Wireless Commun.*, vol. 7, no. 8, pp. 2890–2895, Aug. 2008.

22. A. S. Behbahani, R. Merched, and A. M. Eltawil, "Optimizations of a MIMO relay network," *IEEE Trans. Signal Process.*, vol. 56, no. 10, pp. 5062–5073, Oct. 2008.
23. R. Mo and Y. Chew, "Precoder design for non-regenerative MIMO relay systems," *IEEE Trans. Wireless Commun.*, vol. 8, no. 10, pp. 5041–5049, Oct. 2009.
24. W. Zeng, C. Xiao, M. Wang, and J. Lu, "Linear precoding for finite-alphabet inputs over MIMO fading channels with statistical CSI," *IEEE Trans. Signal Process.*, vol. 60, no. 6, pp. 3134–3148, Jun. 2012.
25. W. Zeng, C. Xiao, J. Lu, and K. B. Letaief, "Globally optimal precoder design with finite-alphabet inputs for cognitive radio networks," *IEEE J. Sel. Aeras Commun.*, vol. 30, no. 10, pp. 1861–1874, Nov. 2012.

Chapter 6
Emerging Research on Information Structures

6.1 Attempting to Find Structures in Images

With reference to Sect. 1.1, Shannon's rate-distortion theorem describes a relationship between information rate and degree of distortion. In order to reduce the degree of distortion, high information rate is usually required. However, bandwidth is often limited in practical wireless multimedia communication systems, which cannot always afford a high-rate transmission for data streams such as real-time videos with acceptable perceptive qualities.

In practice, it is easy to see that pixels in videos and images receive drastically different amounts of attention from the viewers. As information of videos and images is received, what the human mind is concerned with is likely focused upon the intrinsic contents. Interestingly, a recent finding indicates that the human mind perceives information typically based on structural representations [1]. As a matter of fact, the amount of information that the brain needs to process may be very minor. Thus we are confronted with the question: is it possible to represent information from a cognitive view which may help to reduce the required transmission data rate?

For a sample "puma" image shown in Fig. 6.1 with the resolution of 300×198, 6.95 Kb of data was generated after compression using the JPEG2000 standard. How can we find a method to represent this image with a process similar to that of human cognition?

When one recognizes the content of the image, color information may be a kind of a prior semantic knowledge, i.e., white for the background, light brown for the puma, dark brown for the branch, and so on. Accordingly, apart from the color information, texture, shape, and other aspects of the semantic categories may also be used to represent the puma image. Referring to Fig. 2.7b, these representations may also be considered as intermediate structures. It is noted that the structures here are constructed from the viewpoint of human recognition which proceeds from top to bottom. By analyzing the input image based on these structures rather

© The Author(s) 2015
J. Lu et al., *Structural Processing for Wireless Communications*, SpringerBriefs
in Electrical and Computer Engineering, DOI 10.1007/978-3-319-15711-5_6

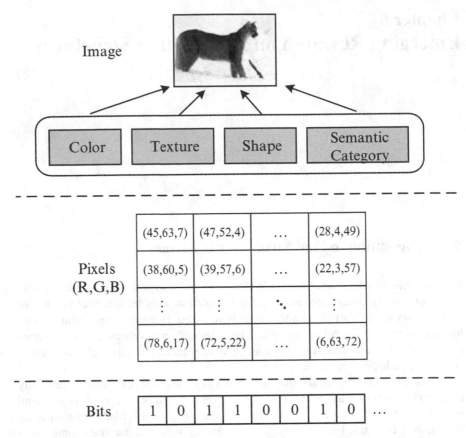

Fig. 6.1 An example "puma" image in multiple levels. The bottom layer is the quantized bits for transmission, and the layer above contains the pixels' values of the image. Between the pixel representation and the image, intermediate units may be discovered for a compact image representation

than pixels, if databases containing different structures and their properties, such as colors, textures and shapes are pre-built, only the labels and parameters of the structures are required to represent the original image. Thus, the amount of data used to characterize the specific image may be significantly reduced compared to those with existing standards.

Assuming that the transmitter and receiver share the pre-built databases before transmission, only the labels and parameters of intermediate structures need to be transmitted, greatly reducing the required transmission bandwidth. As such, by the use of structural representations, some emerging multimedia processing approaches may be incorporated to improve the efficiency of wireless multimedia communications.

6.2 Structural Image Coding with Learned Dictionaries

6.2.1 Learning Basic Structures from Images

As mentioned in Chap. 3, conventional compression methods, such as H.264/ MPEG4 and JPEG2000, mainly utilize orthogonal bases for signal transform to remove spatial correlation inside and between images [2, 3].

All these image and video coding approaches depend on pre-defined signal transform bases, such as discrete wavelet transform (DWT) and discrete cosine transform (DCT) in Fig. 6.2a, while ignoring different properties or structures of different kinds of images. While algorithms for linear transforms may be designed to make hardware distinguish between different amplitudes and frequencies of signals, machines cannot understand the high-level semantic contents of images since they do not have the required structural processing capabilities.

Recently, with the development of neural networks, machine learning, as well as semi-conductor technology, computers are becoming able to conduct training-based algorithms for classification and analysis similar to those realized by the human brain. Hence, it may be possible to learn specially optimized bases to represent specific kinds of images. These bases referred to as learned dictionaries, are learned from large numbers of sample images of the same class with common structures, as shown in Fig. 6.2b. Similar to dictionaries for human languages, a learned dictionary should be designed according to image categories and structures, and such dictionary may be regarded as a set of intermediate structures for the images. In the following parts, we illustrate a case of dictionary design for images.

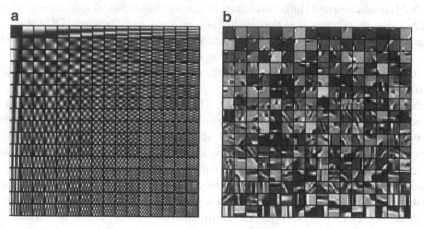

Fig. 6.2 An illustration of a learned dictionary. (a) A complete DCT bases. (b) A set of learned dictionary from some kind of image patches

6.2.2   Image Coding Based on Dictionary Learning

For an efficient image representation model [4], the target of sparse representation with learned dictionary is to describe a signal as the linear combinations of few elements, which are chosen from a large collection of bases. These signal bases may be learned from a variety of image samples, referred to as redundant or over-complete dictionary. Benefitting from sparse representation algorithms, the proposed approach in [5] focuses on the image coding problem using learned dictionaries rather than fixed signal bases. In order to encode image patches effectively, a multi-sample sparse representation (MSR)-based image coding approach is derived with basic elements from collections of natural images. Moreover, the proposed learning algorithm and MSR are incorporated into an image coding framework, which further reduces the reconstructed errors.

As shown in Fig. 6.3, the framework mainly consists of three parts including the MSR-based dictionary learning, MSR-based image encoding, and its decoding counterparts. Before encoding images, the dictionary is trained by the MSR-based dictionary learning algorithm. The MSR-based image coding and decoding are described below.

With reference to some other image compression approaches with block processing, e.g., JPEG, an input image is sliced into non-overlapping image patches over a regular grid. As illustrated in the encoding framework shown in Fig. 6.3, the mean values of image patches and the mean-removed patches $\{\mathbf{x}_j\}_{j=1}^{J}$ are separately encoded, where $\mathbf{x}_j = [x_{1j}, \ldots, x_{nj}]^T \in \mathbb{R}^{n \times 1}$ denotes the vectorized version of an image block. The DC components of an image patch, i.e., the mean value of patch $u_j = \frac{1}{n} \sum_{i=1}^{n} x_{ij}$, is quantized and coded by differential pulse-code modulation (DPCM) prediction and Huffman coding, removing the statistical redundancy. More specifically, it indicates that the residual $e_j = u_j - u_{j-1}$ between DC-values is quantized by $round(\frac{e_j}{scalar})$ and entropy encoded by Huffman coding, where $scalar$ is a pre-defined constant in the quantization step.

Rather than utilizing the DCT or wavelets transform coding, the approach in [5] utilizes MSR to encode the patches $\{\mathbf{x}_j\}_{j=1}^{J}$ to achieve sparse coefficients $\{\boldsymbol{\alpha}_j\}_{j=1}^{J}$ with respect to the learned dictionary $\mathbf{D}$. This dictionary has been trained from a large number of samples chosen from a collection of training images. Moreover, in order to encode the coefficients, the values of non-zero coefficients and the corresponding index of non-zero values are encoded. The non-zero values are quantized and encoded using a Huffman entropy coder, which is similar as the one used for DC values. Then, the indices of the corresponding non-zero coefficients are further coded by fixed length codes using $\log_2 m$ bits, where m denotes the number of dictionary elements. Note that the Huffman table utilized above is constructed off-line and pre-stored at both encoder and decoder sides. This encoding module shares some similarities to the one designed for intra-frame video coding [6, 7].

The compressed image data are mainly composed of the encoded coefficients $\{\hat{\boldsymbol{\alpha}}_j\}_{j=1}^{J}$ and the corresponding DC values in the decoding counterpart. At the decoder side, the compressed image data may be inversely quantized and decoded

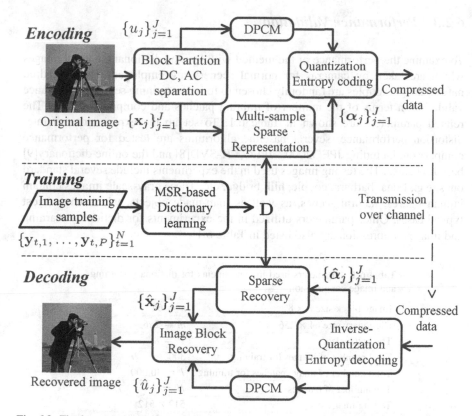

Fig. 6.3 The framework of MSR-based image coding, including the dictionary learning, encoding and its decoding counterparts

according to the Huffman table. The non-overlapped image patches may be recovered with respect to the learned dictionary $\hat{\mathbf{x}}_j = \mathbf{D}\hat{\boldsymbol{\alpha}}_j$. Note that these coded coefficients of patches are followed by the *EOB* (end of block) symbol as in the JPEG standard. Therefore, the total bit rate consumption of a compressed image is calculated as

$$R_{total} = \sum_{j=1}^{J} \{R(\hat{\boldsymbol{\alpha}}_j) + R(\mathrm{DC}_j) + R(\mathrm{EOB}_j)\} + \Delta, \qquad (6.1)$$

where $R(\cdot)$ is the length in bits of the codeword representing the corresponding symbol. Note that the symbol Δ represents the additional bits that include the size of Huffman table, even though the overhead may be ignored.

6.2.3   *Performance Validation*

To examine the performance of the method with learned dictionary, natural images which are taken by cameras for normal scenery are employed. Three-hundred natural image samples are randomly chosen to form the training set for performance validation in terms of the number of training patches and computation time. The related parameters are shown in Table 6.1. To set up the experiments on rate-distortion performance, several baseline algorithms are tested for performance comparison, including JPEG2000,[1] JPEG,[2] K-SVD[8] and the online dictionary[9] based schemes. The testing images used in the experiments include several standard ones, e.g., Lena, barbara, couple, hill, bridge, etc. These gray-scale images contain human figures, natural scenes, as well as man-made objects, which cover most typical cases. Other parameters utilized in the experiments for dictionary learning and image compression are also listed in Table 6.1.

Table 6.1 Parameters used in experiments for dictionary learning and image compression

Image block size 8 × 8	$n = 64$
Dictionary size 64 × 256	$m = 256$
The batch size	$P = 8$
Total number of images for training	300
Total number of image patches for training	$I = 30{,}000$
The number of batches	$N = \frac{I}{P} = 3{,}750$
Testing image size	512 × 512
Number of blocks to encode an image	$J = \frac{512 \times 512}{n}$
Regularization parameter	$\lambda = \frac{C}{\sqrt{m}}, C = 1.2$

Experiments are conducted on these standard gray-scale images to evaluate the rate distortion performance. Figure 6.4 shows the performance comparison between the MSR-based image coding and other methods in terms of PSNR. For typical bit-rate cost, it is shown that there is about 1.98 dB improvement on average over JPEG2000 and 1.04 dB improvement on average over conventional online learning approach. Moreover, compared with the K-SVD dictionary based image compression scheme, the proposed approach is slightly better in the rate distortion performance by 0.61 dB, yet it runs much faster than K-SVD in dictionary training.

From these figures, one may observe that the conventional online dictionary learning based image compression likely induces larger reconstructed errors compared to the globally learned K-SVD dictionary. This validates the disadvantage of online dictionary learning in general. To better fix the problem, MSR is incorporated

---

[1]OpenJPEG, an open-source JPEG2000 codec, http://www.openjpeg.org.

[2]JPEG free library is available at http://www.ijg.org.

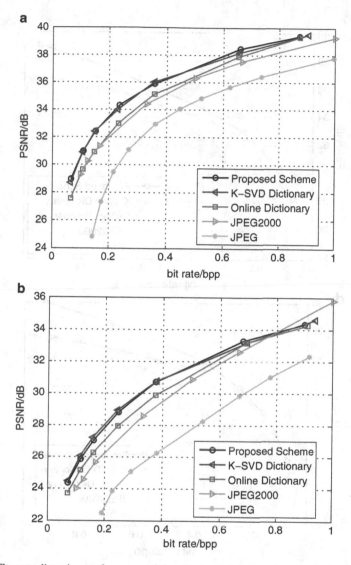

Fig. 6.4 The rate-distortion performance of the proposed algorithm compared with JPEG2000, JPEG, K-SVD and online dictionary based image coding methods in terms of peak signal-to-noise ratio (PSNR) using several test images (size 512×512, gray-level). Note that the *curves with circle markers* show the performance of the proposed MSR-based image coding approach. (**a**) Lena. (**b**) Barbara. (**c**) Hill. (**d**) Boat

into online dictionary to compress natural images, thus achieving much reduced reconstructed errors while maintaining the benefits of online dictionary learning. These results demonstrate the efficiency of the proposed learning algorithm. For subjective quality assessment, the visual results of barbara, boat, hill, etc., are shown

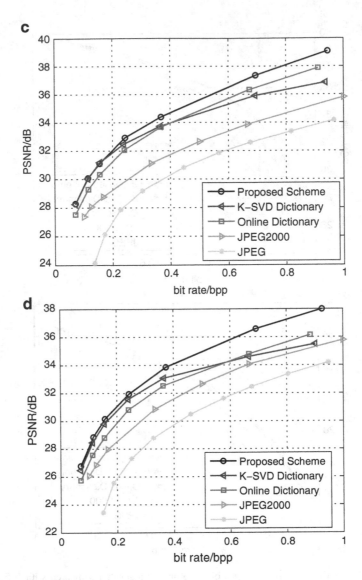

Fig. 6.4 (continued)

in Fig. 6.5. All of these images are encoded at the same bit-rate of 0.25 bpp (bit-per-pixel), and the corresponding PSNR and SSIM [10] values are also listed. From these figures, the recovered images by the MSR-based image coding approach appear more clearly and naturally, and details are much better preserved. In contrast, other methods, e.g., JPEG and JPEG2000, likely induce more blocky or blurring artifacts. This is mainly due to the over-complete dictionary with fewer non-zero coefficients and reduced reconstruction errors that is used for the MSR-based image

Fig. 6.5 Visual results for a subjective quality comparison. All encoded at 0.25 bpp bit-rate (best viewed by zoom-in). The 1st column on the left: original images, the 2nd column: with JPEG, the 3rd column: with JPEG2000 and the last column: with the proposed approach. Note that the PSNR and structural similarity index measure (SSIM) values are listed below the images. (**a**) Barbara, original. (**b**) 24.35 dB, 0.803. (**c**) 27.30 dB, 0.885. (**d**) 28.82 dB, 0.958. (**e**) Boat, original. (**f**) 27.32 dB, 0.845. (**g**) 29.50 dB, 0.895. (**h**) 32.00 dB, 0.971. (**i**) Hill, original. (**j**) 28.61 dB, 0.871. (**k**) 30.01 dB, 0.900. (**l**) 32.84 dB, 0.972. (**m**) Couple, original. (**n**) 27.23 dB, 0.856. (**o**) 28.78 dB, 0.883. (**p**) 31.76 dB, 0.970

coding and dictionary learning approach. The visual quality of images may be further improved at low bit-rate region, and some band limited mobile applications, e.g., mobile image browsing and sharing, may better showcase the advantages of the proposed method.

6.3   Structural Video Coding Based on Learned Models

6.3.1   Model-Based Video Coding

For decades, long-distance face-to-face conversation has been an important aspiration of multimedia communication systems. Unfortunately, its utility is hindered by its stringent delivery requirements including low latency, high reconstruction quality, and the ability to scale well over a channel with highly variable capacity.

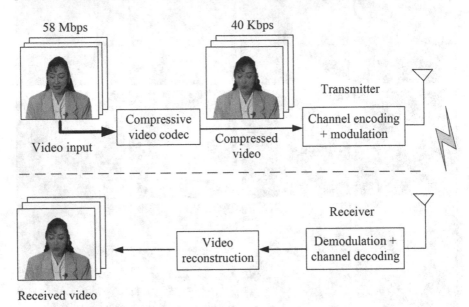

Fig. 6.6 The block diagram of a conventional video transmission system, where the video codec is H.264 and the quality of raw video sequence is 352 × 288@25 fps

Figure 6.6 illustrates the conventional video transmission, where H.264 codec is taken as an example. It can be seen that the images after compression are blurry, because conventional video coding schemes seldom consider the content of the video and regard all pixels on an image as random variables. In other words, the face enjoys the same fidelity as the less important parts in the background. Therefore, high data rate is necessary to support high definition (HD) video streams at professional video conferences. For instance, Cisco® TX9000™ requires at least 8.8 Mbps of transmission data rate to support 1080p@30 HD conversational video under H.264 codec [11]. Hence, applications with real-time conversational video have not yet achieved wide-spread adoption due to the limited system throughput of current wireless networks.

Based on the viewpoint of human cognition, a face video transmission approach that utilizes model-based video coding (MBVC) scheme [12] is introduced in this section. Different from conventional video coding, MBVC makes full use of face structural information to abstract the parameterized variation of each individual face. This coding scheme involving training the computer to recognize the shapes of objects such as eyes, nose, and mouth. The changes of shapes can be then represented by a small number parameters. In addition, the characteristics of these objects can be learned for recognizing new faces. The whole procedure is indeed a mimicry of the face perception function of the human brain. Intuitively, the more training data there are available, the more accurate faces can be represented and reconstructed. Imagine two participants of a video chat sharing the same pair of training set, only labels that indicates the changes on faces need to be sent instead of the whole compressed video, reducing the data rate dramatically. MBVC may represent human faces and other common objects with extremely low bit-rates, which has a potential in wireless multimedia communications.

6.3.2   Learning Structural Information by Face Modeling

It is known that the principle of acquisition in the human brain is making comparison between prior knowledge and the newly arrived information [13]. If someone meets a new face, there is no prior knowledge available. Hence, the brain starts to collect all the structural information on the face including the color of skin, the shape of the nose, the size of eyes, etc., and then to memorize this group of information as database. In computer vision research, it has been shown that by normalizing facial appearances against their shapes, both the shape and the appearance variations of a specific person may be well modeled by linear subspaces [14]. As shown in Fig. 6.7, given an image of a face annotated with the locations of a pre-defined set of landmarks, the shape s of the face is defined as the concatenation of the landmark coordinate values, and the shape-normalized appearance g is obtained by piecewise affine warping of the face image onto a frontal reference shape. Here $\bar{s}$ and $\bar{g}$ may be defined as the reference shape and appearance, respectively, which are usually taken from a frontal neutral face. Performing this procedure on an annotated set of face images results in a shape training set and a shape-normalized appearance training set. From their respective training sets, principal component analysis (PCA) [15] may be used to identify the shape and appearance variations $\mathbf{P_s}$ and $\mathbf{P_g}$ as subspaces, respectively.

Once the linear subspace model has been trained, an annotated face image $\mathbf{I}_0$ with shape $\mathbf{s}_0$ may be warped to form a shape-normalized appearance patch $\mathbf{g}_0$. Both $\mathbf{s}_0$ and $\mathbf{g}_0$ can be projected onto the corresponding subspace to obtain a small set of parameters $\mathbf{b_s}$ and $\mathbf{b_g}$ as follows:

$$\mathbf{b_s} = \mathbf{P_s}^T (\mathbf{s}_0 - \bar{\mathbf{s}})$$
$$\mathbf{b_g} = \mathbf{P_g}^T (\mathbf{g}_0 - \bar{\mathbf{g}}).$$

(6.2)

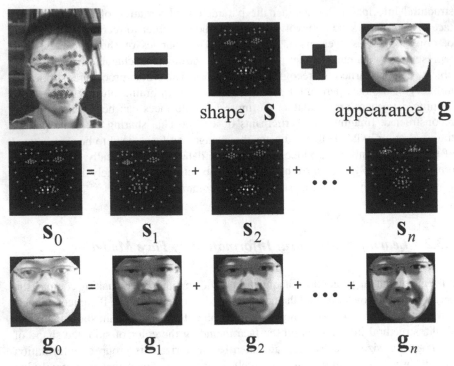

Fig. 6.7 The structural information consisting of shapes and appearances abstracted from a face

Note that b_s and b_g may fully represent the spatio-temporal variation of the face. An approximate facial image reconstruction $\hat{\mathbf{I}}_0$ can be synthesized by the following synthesis equations:

$$\hat{\mathbf{S}}_0 = \bar{\mathbf{s}} + \mathbf{P_s}\mathbf{b_s}$$
$$\hat{\mathbf{g}}_0 = \bar{\mathbf{g}} + \mathbf{P_g}\mathbf{b_g} \qquad (6.3)$$
$$\hat{\mathbf{I}}_0 = W(\hat{\mathbf{g}}_0 : \mathbf{b_s}),$$

where $W(\hat{\mathbf{g}}_0 : \mathbf{b_s})$ denotes the warping operation which warps the shape-normalized texture $\hat{\mathbf{g}}_0$ onto the shape determined by parameters $\mathbf{b_s}$. By representing a face image with a few parameters, subspace models can radically reduce the dimensionality of the representation. They provide the theoretical basis to the utilization of prior structural information on face images in the task of conversational video coding.

6.3.3 MBVC in Video Communications

Since video reconstruction only requires the model parameters, it is only necessary to transmit these parameters instead of the compressed video sequences from the transmitter to the receiver. Note that $\mathbf{P_s}$, $\mathbf{P_g}$, $\mathbf{s_0}$, and $\mathbf{g_0}$ should be shared and stored as parts of face models at both the transmitter and receiver before transmission. This process is illustrated in Fig. 6.8. Clearly, in this case MBVC cannot reconstruct any non-face objects. However, this is a strength instead of a shortcoming: coded face representations are much more compact, and high-quality face video may be maintained even over an unpredictable channel.

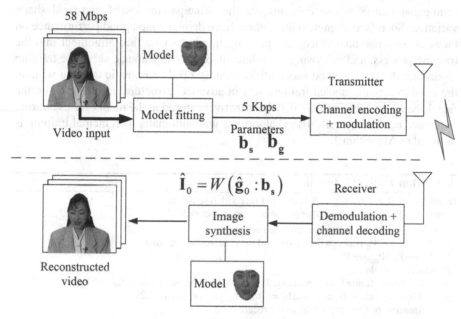

Fig. 6.8 The model-based video communication architecture which only needs to transmit the model parameters

The efficiency of MBVC for conversational video depends critically on the quality of the face model used in the coding process. The typical model training approach outlined in the previous section is unappealing for face MBVC due to its laborious training set collection and annotation process. Works such as [16] have shown that the automatic annotation of landmarks on a given face image (also known as *face alignment*) can be formulated as a regularized constrained optimization problem by attempting to find a most likely configuration of facial landmarks given the face image over all possible facial shapes (See (6.4), where $\|\mathbf{q}\|_\Lambda$ denotes the Mahalanobis norm of the non-rigid parameters $\mathbf{q}$, and Λ denotes the diagonal matrix formed by the eigenvalues of the non-rigid modes of variations according to the shape subspace model).

$$\mathscr{Q}(\mathbf{p}) = \sum_{i=1}^{n} \log(P(l_i = 1 | \mathbf{x}, \mathbf{I})) - \|\mathbf{q}\|_\Lambda, \tag{6.4}$$

where $\mathbf{I}$ is the face image. The positions of the landmarks, denoted by $\mathbf{x}$, are constrained by the shape subspace model as follows:

$$\mathbf{x} = s\mathbf{R}(\bar{\mathbf{x}} + \boldsymbol{\Phi}\mathbf{q}) + \mathbf{t}, \tag{6.5}$$

where $\bar{\mathbf{x}}$ is the mean shape vector of the shape subspace model, and the vector of shape subspace parameters $\mathbf{p} = \{s, \mathbf{R}, \mathbf{q}, \mathbf{t}\}$ consists of scaling factor s, rotation matrix $\mathbf{R}$, translation vector $\mathbf{t}$, and non-rigid parameters $\mathbf{q}$, where $\boldsymbol{\Phi}$ is a column-orthogonal matrix whose columns are the principal modes of non-rigid shape variation. Such face alignment algorithms have demonstrated good performance on faces in conversational video. By incorporating automatic face alignment into the training process, and proposing an online shape and appearance subspace training algorithm, the used method successfully achieved fully automatic training without the need to acquire labeled training sets in advance. Structural information on the facial shapes and appearances may be effectively stored in the model for use during video coding. The overall MBVC algorithm with automatic incremental training is outlined in Algorithm 1.

Algorithm 1 MBVC algorithm

Input: $\mathbf{V} = \{\mathbf{I}_i\}, i \in \{1, \ldots, N\}$: an input face video sequence
Input: $\mathbf{P_s}$, $\mathbf{P_g}$: linear shape and appearance subspace models
Output: $\hat{\mathbf{V}} = \{\hat{\mathbf{I}}\}$: the reconstructed video sequence
1: Share $\mathbf{P_s}$ and $\mathbf{P_g}$ between the sender and the receiver before hand.
2: $\mathbf{P_{s0}} \leftarrow \mathbf{P_s}$, $\mathbf{P_{g0}} \leftarrow \mathbf{P_g}$.
3: **while** $i < N$ **do**
4:     Use the ith frame $\mathbf{I}_i$ to incrementally update the subspace models $\mathbf{P_{s0}}$ and $\mathbf{P_{g0}}$.
5:     Align the face in $\mathbf{I}_i$, compute $\mathbf{b}_i = [\mathbf{b}_{si}; \mathbf{b}_{gi}]$ according to (6.2).
6:     Transmit $\mathbf{b}_i$ from the sender to the receiver.
7:     Reconstruct $\hat{\mathbf{I}}_i$ at the receiver according to (6.3).
8: **end while**
9: **if** $\mathbf{P_{s0}}$ and $\mathbf{P_{g0}}$ give better performance than $\mathbf{P_s}$ and $\mathbf{P_g}$ **then**
10:     $\mathbf{P_s} \leftarrow \mathbf{P_{s0}}$, $\mathbf{P_g} \leftarrow \mathbf{P_{g0}}$.
11: **end if**
12: **return** $\{\hat{\mathbf{V}}\}$

To evaluate the above coding method, it is applied to two 25 fps CIF (352 × 288) conversational video sequences, *akiyo*[3] and *Franck*.[4] These experiments aim to study the performance characteristics of the MBVC method by comparing the

---

[3] http://trace.eas.asu.edu/yuv/akiyo/akiyo_cif.7z.
[4] http://www-prima.inrialpes.fr/FGnet/data/01-TalkingFace/talking_face.html.

results to generic high efficiency video coding (HEVC), region-of-interest (ROI) based HEVC (as proposed in [17]), as well as mesh-based coding (as proposed in [18], with HEVC-coded residuals).

In addition to the coding of faces, the overall conversational video coding system must also code the rest of the visible parts of the body, as well as the background. Inspired by advances in specialized segmentation algorithms such as [19], an automatic head-shoulder segmentation algorithm is implemented here based on the face location reported by the face coding step.

After the segmentation, the background may often be assumed static; thus an extremely low frame-rate (even a still image) suffices for the background. Since the viewer's sensitivity to reconstruction errors in the rest of the visible body parts is relatively low, more distortion can be tolerated, leading to a significantly lower overall bit-rate. In summary, the coding system here consists of three coded video layers:

- The face layer, coded with MBVC to maintain high quality at low bit-rate by utilizing extensive prior structural information;
- The visible body parts layer, coded with a generic video codec on low-quality setting; and
- The background layer, coded with either (a) a generic video codec at a very low frame-rate, or (b) substituted by a still image.

Experiments and real-life application have shown that this layer-wise coding of conversational video consistently yields a 50–70 % reduction in required bit-rate for a given subjective quality level over conventional generic video codes.

Due to the real-time nature of our application, B-frames were disabled in all HEVC-coded sequences in our experiments; the $x265$ implementation[5] of the HEVC encoder is capable of achieving real-time encoding with the "medium" preset on PCs with Intel® Core™ i7 Processors. The Rate-Distortion (R-D) characteristics of the four coding methods on each of the testing sequences in terms of the PSNR are shown in Fig. 6.9. In addition to overall R-D, R-D curves for the face region alone are also shown. Figure 6.10 shows still frames from the test sequences, encoded with all four methods at similar bit-rates.

It is observed that, at low bit-rates, the MBVC method exhibits superior R-D performance, especially for the face region. The face region R-D curves for both mesh-based and generic HEVC are actually below their respective overall R-D curves, implying that they coded the face region at a lower quality than the rest of the video, likely due to its high-frequency details and frequent movements. The ROI-based HEVC improves face region performance for *Akiyo* at the cost of reduced non-face reconstruction quality; however, for *Franck*, its effectiveness is very limited since the talker is much closer to the camera; consequently, the resultant ROI constitutes a large portion of the screen. By utilizing prior structural information, MBVC may achieve greatly improved performance in the face region, thus

---

[5]http://x265.org/.

enhancing perceived quality. It should be noted that due to inevitable unmodeled dynamics and noise in the video sequence, the gain of MBVC diminishes as bit-rate increases; at high bit-rates, it may even become negative compared to HEVC owing to its additional transmission of model parameters and mesh vertex motion vectors.

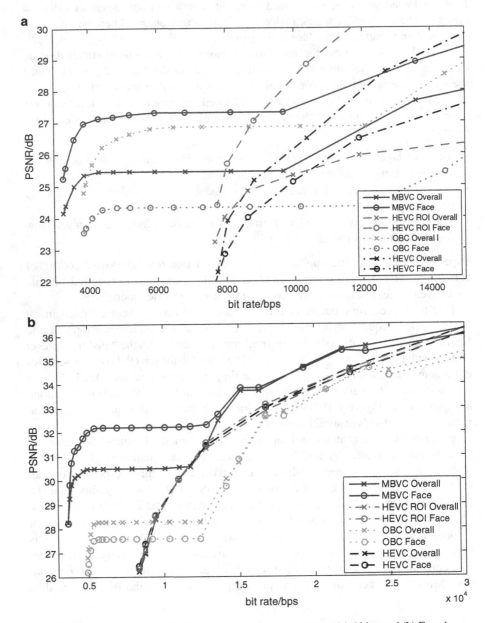

Fig. 6.9 The comparisons for the R-D curves on testing sequences (**a**) Akiyo and (**b**) Franck

A simple codec selection mechanism may be designed to switch between MBVC and HEVC to ensure optimal performance across all bit-rates.

Fig. 6.10 The subjective quality comparisons of the codecs on testing sequences (**a**) Akiyo and (**b**) Franck

6.4 Computational Communication Architecture Based on Structural Processing

In previous sections, cognition-oriented information representations have been shown to exhibit better perceptive quality than conventional ones especially at low transmission data rates. Furthermore, if one can exploit the structural information to obtain a highly efficient representation, the amount of data to be transmitted in wireless communication systems should be much reduced.

Stimulated by the ideas in Sects. 6.2 and 6.3, one may establish an innovative computational communication architecture for wireless communications based on structural processing, as is shown in Fig. 6.11. To incorporate structural information, a *structure base* is built and shared at both transmitter and receiver as the prior information to represent images or videos likely as those stored in human brains. By the use of strong computing capability in networks, the input information is expressed by the prior structural information units stored in the structure base at the transmitter and is reconstructed by the aid of corresponding units at the receiver. Based on this architecture, the data transmitted via the wireless channels are only the indexes and parameters of the structural units, and the transmission efficiency of wireless communications may be dramatically improved.

Assume that the messages at the transmitter and the receiver are expressed as X and Y, respectively. According to information theory, the minimal information to be transmitted via channel should be equal to the mutual information between X and Y, i.e., $I(X, Y)$. However, if a structure base of prior structural information used for signal representation is pre-stored at both ends, the information to be transmitted would be a conditional mutual information M which exhibits a property as:

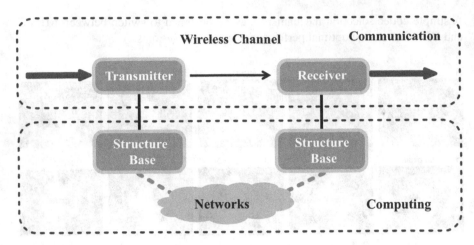

Fig. 6.11 A diagram of the computational communication architecture based on structural processing

$$M \ll I(X, Y). \tag{6.6}$$

As a result, the bandwidth for transmission may by effectively saved.

Thanks to the idea with the structural priori knowledge, one may make a breakthrough to the bottleneck of conventional methods, and open up a new trend for the development of processing technologies. In future wireless communications, such a computational communication architecture may have a great potential to be applicable in enhancing transmission efficiency, effectively solving the seemingly insurmountable problems of complexity and uncertainty.

6.5   Conclusions

In this chapter, in order to find efficient methods of information representations for multimedia, dictionary learning and model-based coding methods are both introduced to abstract prior structural information such that the efficiency of wireless multimedia communications may be significantly improved. Based on such emerging research on structural information, a computational communication model is further introduced to provide a feasible structural processing architecture greatly reduced transmission bandwidth requirement. Further research on structural processing technology is a promising direction, from which a new theory for modern wireless communications may eventually emerge.

References

1. J. B. Tenenbaum, C. Kemp, T. L. Griffiths, and N. D. Goodman, "How to grow a mind: Statistics, structure, and abstraction," *Science*, vol. 331, no. 6022, pp. 1279–1285, 2011.
2. D. Marpe, "The H.264/MPEG4 advanced video coding standard and its applications," in *IEEE Communications Magazine*, vol. 44, no. 8, pp. 134–143, Aug. 2006.
3. ISO/IEC 15444-1, "JPEG 2000 Part I Final Committee Draft Version 1.0", 2000.
4. R. Rubinstein, "Dictionaries for Sparse Representation Modeling," *Proc. IEEE*, vol. 98, no. 6, pp. 1045–1057, Jun. 2010.
5. Y. Sun, X. Tao, Y. Li, and J. Lu, "Dictionary learning for image coding based on multi-sample sparse representation," *IEEE Trans. Circ. Syst. Video Technol.*, vol. 24, no. 11, pp. 2004–2010, Apr. 2014.
6. Y. Sun, M. Xu, X. Tao, and J. Lu, "Online dictionary learning based intra-frame video coding via sparse representations," in *Proc. WPMC'12*, pp. 16–20, Sep. 2012.
7. Y. Sun, M. Xu, X. Tao, and J. Lu, "Online Dictionary Learning Based Intra-frame Video Coding," *Springer Wireless Personal Communications*, vol. 74, no. 4, pp. 1281–1295, 2014.
8. M. Aharon, and M. Elad, "K-SVD: An Algorithm for Designing Overcomplete Dictionaries for Sparse Representation," *IEEE Trans. Signal Process.*, vol. 54, no. 11, pp. 4311–4322, Nov. 2006.
9. J. Mairal, and F. Bach, "Online Learning for Matrix Factorization and Sparse Coding," *Journal of Machine Learning Research*, vol. 11, pp. 19–60, 2010.
10. Z. Wang, A. C. Bovik, H. R. Sheikh, and E. P. Simoncelli, "Image quality assessment: From error visibility to structural similarity," *IEEE Trans. Image Process.*, vol. 13, no. 4, pp. 600–612, 2004.
11. Cisco Systems, *Cisco TelePresence Product Catalog*, 2013. [Online]. Available: http://www.cisco.com/c/en/us/products/collaboration-endpoints/telepresence-tx9000-series/brochure-listing.html. [Accessed Apr. 15, 2014].
12. Aizawa, Kiyoharu and Huang, Thomas S, "Model-based image coding advanced video coding techniques for very low bit-rate applications," *Proceedings of the IEEE*, vol. 83, no. 2, pp. 259–271, 1995.
13. J. D. Weiland, M. S. Humayun, "Visual prosthesis," *Proc. IEEE,* vol. 96, no 7, pp. 1076–1084, July 2008.
14. R. Gross, I. Matthews, and S. Baker, "Generic vs. person specific active appearance models," *Image Vis. Comput.*, vol. 23, no. 12, pp. 1080–1093, Nov. 2005.
15. I. T. Jolliffe, *Principal Component Analysis, Series: Springer Series in Statistics*, 2nd ed., Springer, NY, 2002.
16. J. M. Saragih, S. Lucey, and J. F. Cohn, "Deformable Model Fitting by Regularized Landmark Mean-Shift," *Int. J. Comput. Vis.*, vol. 91, no. 2, pp. 200–215, Sep. 2011.
17. L. S. Karlsson, and M. Sjostrom, "Improved ROI video coding using variable Gaussian pre-filters and variance in intensity," in *Proc. IEEE ICIP'05*, Genoa, Italy, pp. 313–316, Sep. 2005.
18. P. van Beek, A. M. Tekalp, N. Zhuang, and I. Celasun, "Hierarchical 2-D Mesh Representation, Tracking, and Compression for Object-based Video," in *IEEE Trans. Circ. Syst. Video Technol.*, vol. 9, no. 2, pp. 353–369, Mar. 1999.
19. H. Xin, H. Ai, H. Chao, and D. Tretter, "Human Head-shoulder Segmentation," in *Proc. IEEE FG'11*, Santa Barbara, USA, pp. 227–232, Mar. 2011.